PROBLEMS IN PHYSICS

for Advanced Level and Scholarship Candidates
(SI Version)

Other Physics texts by
Dr. F. Tyler

A- and S- level

A Laboratory Manual of Physics
Heat and Thermodynamics

PROBLEMS IN PHYSICS

For Advanced Level and Scholarship Candidates

(SI version)

F. TYLER

B.Sc., Ph.D., F.Inst.P.

Formerly Senior Science Master
Queen Elizabeth's Grammar School, Blackburn

Edward Arnold

First published 1957
by Edward Arnold (Publishers) Ltd
41 Bedford Square
London, WC1B 3DQ
Reprinted 1959, 1962, 1964, 1967
SI Edition 1971
Reprinted (with corrections) 1973, 1977, 1979

ISBN: 0 7131 2295 1

To D

Printed in Great Britain
by Unwin Brothers Limited
The Gresham Press, Old Woking, Surrey, England
A member of the Staples Printing Group

PREFACE

This present volume sees a complete overhaul of the first edition of my problems book to service it for the 'SI' era. This has been done to the exact specifications set out in the BSI publications on this topic and, especially, in conformity with the excellent report put out by the Association for Science Education.

A large number of the earlier problems have been removed altogether, whilst those remaining have been re-stated and re-evaluated in the new unitary system. The excision process has enabled me to include a wide range of new problems—almost 200 in all—covering all sections of the A- and S-syllabuses but with especial emphasis on dynamics, electrostatics, electro-magnetism and modern physics. Also included for the first time are a number of problems on rocketry and satellite motion. A complete set of answers for all the problems is provided together with a continued sequence of worked problems, whilst at the appropriate stages throughout the text will be found tables of formulae, quantities and units relevant to the sections that follow.

In all there are over 650 varied problems here for all stages of sixth form work and it is the author's hope that they will provide a fully adequate selection of examples for teachers and students in their course preparation.

Blackburn 1971 F. Tyler

CONTENTS

MECHANICS AND HYDROSTATICS

Statics 1
  Worked example 1
Friction 3
  Worked example 3
Hydrostatics 5
  Worked example 5
  Worked example 7
Dynamics 8
  Quantities 8
  Some relations between linear and rotational quantities 9
  Some moments of inertia 9
  Moment of inertia theorems 9
Linear Dynamics 9
  Worked example 10
  Worked example 13
  Worked example 15
Uniform circular motion 18
  Worked example 18
Simple harmonic motion 20
  Worked example 20
Rotational dynamics 24
  Worked example 24
  Worked example 28

PROPERTIES OF MATTER

Elasticity 31
  Worked example 31
Surface tension 33
  Worked example 34
  Worked example 36
Viscosity 38
  Worked example 39
Gravitation 41
  Worked example 42
Rocketry and satellite motion 43
Method of dimensions 45
  Worked example 45

HEAT

Heat—some data and useful constants 48
Thermometry 48
Expansion of solids 49
  Worked example 50
Expansion of liquids 51
  Worked example 52
Expansion of gases. The gas laws 54
  Worked example 55
  Worked example 57
Specific heat capacity 58
  Worked example 59
Latent heats 61
  Worked example 62
Specific heat capacities of gases 64
  Worked example 64
  Worked example 66
Unsaturated vapours and vapour pressure 68
  Worked example 69
Kinetic theory of gases 71
  Worked example 72
Conductivity 74
  Worked example 75
  Worked example 77
Radiation 79

LIGHT

Reflection at plane surfaces 80
Reflection at curved surfaces 80
  Worked example 81
Refraction at plane surfaces. Prisms 82
  Worked example 84
Refraction at curved surfaces. Lenses 85
  Worked example 86
  Worked example 88
Dispersion by prisms and lenses 92
  Worked example 92
Rainbows 94
The eye. Defects of vision 95
  Worked example 95

Optical instruments 96
Worked example 97
Velocity of light 99
Photometry 100
Worked example 101
Wave theory 103
Worked example 105

SOUND

Sound waves. Characteristics of musical sound 108
Worked example 109
Velocity of sound 110
Worked example 110
Vibrations in gas columns 112
Worked problem 112
Worked example 114
Worked example 115
Vibrations in strings 117
Worked example 118
Worked example 120
Doppler effect 121
Worked example 122
Doppler effect in light 123

ELECTROSTATICS

Some formulae and units 124
Forces between charges. Potential 125
Worked example 125
Capacitors. Energy of charge 127
Worked example 128
Electrostatic instruments. Forces on charged bodies 132
Worked example 132

CURRENT ELECTRICITY

Ohm's law and resistance. Kirchoff's laws 135
Worked example 135
Worked example 138

Electrical measurements. The potentiometer and Wheatstone bridge 140
Worked problem 141
Thermo-electricity 144
The chemical effect of a current. Cells 145
Worked example 145
The heating effect of a current. Electrical energy and power 147
Worked example 147
Worked example 149
Electro-magnetism 150
Some formulae and units 150
Electro-magnetism. Galvanometers 150
Worked example 152
Electromagnetic induction. Inductance. Motors and dynamos 157
Worked example 158
Worked example 160
Worked example 161
Worked example 163
Alternating current 165
Some fundamental properties of a.c. circuits 165
Alternating current circuits 165
Worked example 167
Atomic physics. Electronic and modern physics 171
Some data and useful constants 171
Worked example 172
Worked example 173
Worked example 175

ANSWERS TO NUMERICAL QUESTIONS

Mechanics and hydrostatics 180
Properties of matter 181
Heat 181
Light 182
Sound 183
Electrostatics 183
Current electricity 184
Tables of logarithms 188

MECHANICS AND HYDROSTATICS

Statics

1 The top of a circular table, of diameter 2 m, of mass 20 kg, is supported by three legs each of mass 5 kg. The legs are placed near the circular edge of the table so as to form an equilateral triangle. What is the minimum mass, placed on the table top, which will overturn the table?

2 Define *centre of gravity* of a body.

A thin circular metal sheet, of diameter 10 cm, has a smaller circular section of diameter 5 cm removed from one side of the centre and transferred to a symmetrical position on the disc on the opposite side of the centre. Where is the centre of gravity of the final arrangement?

Worked example

3 *To the edge of a square of sheet metal with side 6 cm is attached the base of an isosceles triangular piece of the same material. If this triangle has a base of 6 cm and a vertical height of 6 cm, calculate the position of the centre of gravity of the composite sheet.*

The C.G. of the square piece is at the point of intersection of the diagonals, and the C.G. of the triangular piece one-third the way up the median measured from the base. These two points are marked G_1 and G_2 respectively on the diagram. If, now, the superficial density of

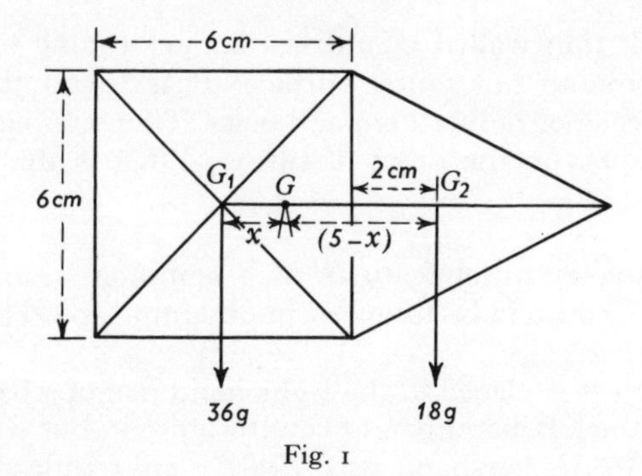

Fig. 1

the sheet metal is w g cm^{-2}, the masses of the two portions are $36w$ g and $\frac{1}{2} \times 6 \times 6w = 18w$ g respectively acting at their corresponding C.G.s.

The problem of finding the C.G. of the composite sheet now becomes that of determining where the resultant of the forces on the above masses act. Let this point be denoted by G distant x from G_1 on the line joining G_1 to G_2. Then, taking moments about G,

$$36w \times x = 18w \times (5-x)$$

On cancelling this gives

$$2x = 5-x$$

from which

$$x = \tfrac{5}{3} = \underline{1{\cdot}66 \text{ cm}}$$

thus specifying the position of G as described.

4 A metre-long piece of stout copper wire has a right-angled bend made in it one-third the distance from one end. The L-shaped wire thus formed is now supported at the right-angled bend by a smooth nail. What position will the wire assume?

5 State fully the conditions of equilibrium for a rigid body acted on by a number of forces.

Two light rigid uniform rods, each 2·5 m long, are freely hinged together at their upper ends, their lower ends resting on a smooth horizontal surface and being connected together by a rope 4 m long. A mass of 10 kg is attached to the mid-point of the left-hand rod, whilst a mass of 30 kg is attached to the mid-point of the right-hand rod.

Calculate the reactions at the floor and at the hinge and the tension in the connecting rope.

6 A length of thin-walled cylindrical copper tubing of radius 3 cm stands on a smooth horizontal surface. It is found that when two ball bearings, each of radius 2 cm and mass 200 g, are placed inside the cylinder, it is just on the point of tilting. What is the weight of the cylinder?

7 Describe the essential features of a common beam balance and outline the procedure to be followed in obtaining an accurate weighing with such a balance.

When an object is placed in the right-hand pan of a balance, 25·00 g are needed in the left-hand pan to counterpoise it, but when the object is transferred to the left-hand pan, 24·01 g are required in the right-hand pan to re-establish counterpoise. Calculate (*a*) the true mass of the object, (*b*) the ratio of the lengths of the balance arms.

8 The beam of a common balance is horizontal with 25·62 g in the left-hand pan and 25·50 g in the right-hand pan, and again with 18·87 g in the left-hand pan and 18·75 g in the right-hand pan. What information does this data give about the state of the balance?

9 What do you understand by the *sensitivity of a balance*? On what factors does the sensitivity depend? Describe how you would determine the sensitivity experimentally.

The three knife edges of a beam balance are in a straight line, the outer knife edges each being 10 cm from the centre one. It is found that when a centigramme mass is placed in one pan of this balance, the end of the 15-cm long pointer moves through a distance of 2·5 mm. Find the position of the centre of gravity of the beam if the mass of the beam and pointer is 30 g.

10 Define the terms *mechanical advantage, velocity ratio* and *efficiency* in reference to machines and establish the relationship between these three quantities.

In a block and tackle with four pulleys in each block an effort of 45·5 kg is needed to steadily raise a load of 295 g. When the effort is reduced to 34·5 kg, the load just begins to run back. Find (*a*) how much of the effort is used in overcoming friction, (*b*) the weight of the lower pulley block.

11 Describe a machine which depends on the screw principle and obtain an expression for the mechanical advantage assuming it to be perfect.

In a screw press a force of 20 N is applied tangentially at each end of the handle whose total length is 20 cm. If the pressure plate has an area of 40 cm^2 and the pitch of the screw is 0·25 cm, what is the pressure transmitted by the plate assuming the press to be 60 per cent efficient?

12 Define *coefficient of statical friction.*

A block of wood is just on the point of slipping when the board on which it is placed is tilted at an angle of 30° to the horizontal. If the block has a mass of 100 g, calculate the limiting frictional force acting and the coefficient of statical friction for the two surfaces.

Worked example

13 *A ladder,* 9 *m in length, rests at an angle of* 60° *to the horizontal against a smooth vertical wall. Calculate the frictional force and the total* *reaction of the ground if the ladder has a mass of 25 kg and its centre of gravity is one-third the way up from the bottom of the ladder.* ($g = 9{\cdot}8\ m\ s^{-2}$).

The forces acting on the ladder are as shown in the diagram. N_1 is the normal reaction of the wall where the ladder touches it at A, and N_2 the normal reaction of the ground where the end of the ladder rests

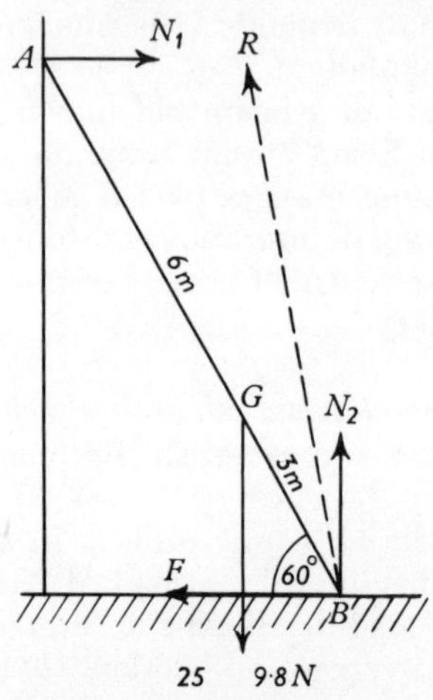

Fig. 2

at B. (From consideration of the vertical equilibrium of the ladder it is clear that the upward force N_2 = downward pull of gravity (weight) on the ladder.) F is the frictional force acting at B to prevent slipping.

Taking moments about A, and expressing all forces in Newtons, we have

$$F \times 9 \cos 30 + 25 \times 9{\cdot}8 \times 6 \sin 30 = N_2 \times 9 \sin 30$$
$$= 25 \times 9{\cdot}8 \times 9 \sin 30$$

Hence

$$F = \frac{(225 - 150)\ 9{\cdot}8 \sin 30}{9 \cos 30}$$

$$= \frac{25 \times 9{\cdot}8}{3} \tan 30 = \underline{47{\cdot}04}\ \text{N}$$

The total reaction (R) of the ground (shown in dotted line in the diagram) is the resultant of the forces F and N_2 at B.
Thus

$$R = \sqrt{F^2 + N_2^2}$$
$$= \sqrt{(47{\cdot}04)^2 + (245)^2} = \underline{249{\cdot}5}\ \text{N}$$

14 If, in the above question, the coefficient of friction between the ladder and the ground is 0·25, find how far a man of mass 75 kg can climb up the ladder without it slipping from under him.

15 Define: *limiting friction, angle of friction.*

You are provided with a plane having a variable angle of slope and a suitable wooden block. Describe how you would use this equipment to obtain a value for the coefficient of limiting friction between the block and the surface of the plane.

In such an experiment the block, which has a mass of 100 g is just on the point of sliding down the plane when the angle of slope is 30°. What is the resultant force acting on the block down the plane when the angle of slope is increased to 60°?

16 A tension of 0·01 N applied at one end of a rope coiled three times round a cylindrical post is found to balance a tension of 5 N at the other end. What is the coefficient of friction between the rope and the post? Establish any formula used in your calculation.

Hydrostatics

17 Give details of the experimental procedure you would adopt to find the density of (*a*) a small irregular piece of resin, (*b*) powdered resin.

A relative density bottle has a mass of 27·07 g when empty, 35·75 g when a certain amount of salt is placed in it, and 75·56 g when the bottle is topped up with turpentine. When filled respectively with turpentine and with water the mass of the bottle and contents is 70·29 g and 76·98 g. Calculate a value for the relative density of salt.

Worked example

18 *Concentrated sulphuric acid of relative density* 1·8 *is mixed with water in the proportion of one to three by volume. If the resulting mixture has a relative density of* 1·28, *what percentage contraction takes place on mixing?*

Let 1 volume of conc. sulphuric acid + 3 volumes of water become x volumes on mixing. Now 1 volume of conc. sulphuric acid weighs 1·8 × weight of 1 volume of water. Hence, total weight of mixture = (1·8 + 3) × weight of 1 volume of water. But the relative density of the mixture is 1·28.

∴ x volumes of the mixture weigh 1·28 × weight of 1 volume of water.

i.e. $1{\cdot}28x = 4{\cdot}8$

or $x = \dfrac{4{\cdot}8}{1{\cdot}28} = 3{\cdot}75$ volumes

Hence contraction on mixing

$$= \left(\frac{4-3{\cdot}75}{4}\right) \times \frac{100}{1} = \underline{6{\cdot}25} \text{ per cent}$$

19 A cylinder of wood floats with its axis vertical in a liquid with three-quarters of its length submerged. The liquid is now diluted with half its volume of water when the cylinder floats with four-fifths of its length submerged. Find the relative density of the liquid and the wood. (Assume that the liquid is freely miscible with water and that no chemical reaction takes place.)

20 A uniform cylindrical rod of wood 20 cm long floats upright with 4 cm of its length above the free surface of a 10 cm deep oil layer standing on the surface of water contained in a deep cylinder. If the relative density of the oil is 0·8, what is that of the wood?

21 The vertical height of a right circular cone is 10 cm and its base, which has a diameter of 10 cm, is covered with a sheet of lead 1 mm thick. If, on being placed in water, the loaded cone floats with 2 cm of its height projecting above the surface, what is the relative density of the wood? Take the relative density of lead as 11·4.

22 Explain what is meant when the weight of a body is given *in vacuo*. How are such values obtained?

Using a beam balance a body of density 2·40 g cm^{-3} is found to weigh 0·6234 N at 10°C and 740 mm when using brass weights. Calculate the weight of the body *in vacuo* if the density of brass is $8{\cdot}45 \times 10^3$ kg m^{-3} and that of air at s.t.p. is $1{\cdot}293 \times 10^{-3}$ kg m^{-3}. (Standard atmospheric pressure $= 1{\cdot}01325 \times 10^5$ N m^{-2} = 760 mm of mercury.)

23 Give the details of the determination of the relative density of (*a*) a solid, (*b*) a liquid by a method involving the use of Archimedes' principle.

A certain solid weighs 125·0 N in air, 75·0 N when fully immersed in a liquid A, and 62·5 N when similarly immersed in a liquid B. If, in a mixture of the liquids, it weighs 70·0 N, what is the proportion by volume of each liquid in the mixture? (Assume no chemical action takes place on mixing the two liquids.)

24 Define *centre of buoyancy, metacentre* and *metacentric height* of a floating body. Describe how the metacentric height of a ship could be determined and discuss the significance of this measurement from the point of view of the ship's stability.

25 What do you understand by the 'law of flotation'?
A block of wood of mass 200 g and density 0·8 g cm^{-3} floats in a liquid of relative density 1·2. What mass of copper, density 8·9 g cm^{-3} must be attached to the wood in order just to sink the combination?

26 Two pieces of aluminium (relative density 2·7), A and B, are suspended from the scale pans of a balance. When A is completely immersed in water and B in paraffin (relative density 0·8), the balance is exactly counterpoised, but when A is immersed in paraffin and B in water, an additional 10 g in B's scale pan is required to re-establish counterpoise. Find the masses of A and B.

Worked example

27 *A hydrometer has a mass of* 20 g *and the area of cross-section of its stem is* 0·25 cm^2. *Calculate (a) the distance between the* 0·80 *and* 1·00 *markings on its stem, and (b) the density reading corresponding to a point mid-way between these two marks.*

When floating in a liquid the hydrometer will displace its own mass (viz. 20 g) of that liquid. Hence, taking the density of water as 1 g cm^{-3}, we see that the volume of the hydrometer below the 1·00 marking must be 20 cm^3.

(*a*) Let the distance between the 1·00 and 0·80 markings on the stem be l cm. Then, when the hydrometer is floating in a liquid at the 0·80 mark, the volume of the liquid displaced will be $(20+0{\cdot}25\,l)$ cm^3 and the mass of this volume of displaced liquid must be 20 g.

Hence $(20+0{\cdot}25\,l)\times 0{\cdot}8 = 20$

or $0{\cdot}2\,l = 20-20\times 0{\cdot}8$

from which $l = \underline{20\text{ cm}}$

(*b*) When floating mid-way between the 1·00 and 0·80 marks, the volume of liquid displaced by the hydrometer

$$= (20+0{\cdot}25\times 10) = 22{\cdot}5\text{ cm}^3$$

If this liquid has a density of x g cm^{-3}, we have

$$22{\cdot}5x = 20$$

$$\text{i.e.}\qquad x = \underline{0{\cdot}89}$$

28 Describe Nicholson's hydrometer and explain how you would use it to find the relative density of a liquid.

A mass of 75·2 g is required to sink a given Nicholson's hydrometer to a certain mark on the stem when immersed in water. With a solid placed first on the upper pan and then on the lower pan the masses needed to sink the hydrometer to the same mark in water are respectively 48·4 g and 59·2 g. What is the relative density of the solid?

29 Explain the meaning of the terms *centre of pressure* and *resultant thrust* as used in hydrostatics.

A lock gate 5 m wide and 6 m deep has water to a depth of 4·5 m and 1·5 m respectively on the two sides. Calculate the resultant thrust on the gate and the point at which it acts. (Density of water = 10^3 kg m^{-3}.)

30 Define *pressure at a point* in a fluid.

A rectangular tank $20 \times 5 \times 4$ cm is completely filled with a liquid of density $1{\cdot}2 \times 10^3$ kg m^{-3}. What is the thrust exerted by the liquid on the end, base and side of the tank?

Dynamics

Quantities:

Linear		*units*
Displacement	$= s$	m
Velocity	$= v$	m s^{-1}
Acceleration	$= a$	m s^{-2}
Mass	$= m$	kg
Force	$= F\ (= ma)$	N
Energy/Work	$= E/W$	J
Kinetic energy	$= \frac{1}{2} mv^2$	J
Momentum $= p$	$= mv$	N s

Rotational		*units*
Angular displacement	$= \theta$	rad
Angular velocity	$= \omega$	rad s^{-1}
Angular acceleration	$= \alpha$	rad s^{-2}
Moment of inertia	$= I$	kg m^2
Torque	$= T\ (= I\alpha)$	N m
Rotational energy/work	$= T\theta$	N m rad
Rotational kinetic energy	$= \frac{1}{2} I\omega^2$	J
Angular momentum	$= L\ (= I\omega)$	J s

Equations of motion:

Linear	*Rotational*
$v = u + at$	$\omega_2 = \omega_1 + \alpha t$
$s = ut + \frac{1}{2}at^2$	$\theta = \omega_1 t + \frac{1}{2}\alpha t^2$
$v^2 = u^2 + 2as$	$\omega_2^2 = \omega_1^2 + 2\alpha\theta$

Some relations between linear and rotational quantities:

Angular velocity $= \omega = \dfrac{v}{r}$ rad s^{-1}

Centripetal acceleration $= a = \omega^2 r = \dfrac{v^2}{r}$ m s^{-2}

Centripetal force $= F = m\omega^2 r = \dfrac{mv^2}{r}$ N

Cycle $= 2\pi$ rad

Frequency $= f = \dfrac{\omega}{2\pi}$ s^{-1} or Hz

Periodic time $= T = \dfrac{2\pi}{\omega}$ s

Some moments of inertia

Moment of inertia $= I = m \times (\text{radius of gyration})^2$

—of thin ring or thin-walled cylinder rotating about its central axis
$= mr^2$ kg m^2

—of uniform disc or solid cylinder rotating about its central axis
$= mr^2/_2$ kg m^2

—of solid sphere rotating about a diametral axis
$= \frac{2}{5}mr^2$ kg m^2

—of hollow sphere rotating about a diametral axis
$= \frac{2}{3}mr^2$ kg m^2

—of uniform rod rotating about a perpendicular axis through C.G.
$= m\left(\dfrac{L^2 + B^2}{12}\right)$ kg m^2

Moment of inertia theorems:

(*a*) Theorem of perpendicular axis,
$I_Z = I_X + I_Y$ kg m^2

(*b*) Theorem of parallel axis
$I = I_G + mh^2$ kg m^2

Linear dynamics

31 Define *uniform acceleration.* Distinguish between the types of motion in which a moving body is subject (*a*) to a constant force, (*b*) to no resultant force. Give examples of each.

Starting from rest, a car travels for 1 minute with a uniform acceleration of 1 m s^{-2} after which the speed is kept constant until the car is finally brought to rest with a retardation of 2 m s^{-2}. If the total distance covered is 4500 m, what is the time taken for the journey?

32 A body is placed on a rough inclined plane and will *just* slide down the plane when the inclination of the sloping surface is 30° with the horizontal. If, now, the angle of slope is increased to 60°, with what acceleration will the body slide down the plane?

33 Describe the motion of a body falling freely under gravity and establish a relationship between the distance covered and the time elapsing after being released from rest.

A balloon, ascending with a steady vertical velocity of 10 m s^{-1} releases a sandbag which reaches the ground 15 seconds later. Neglecting friction, find the height of the balloon when the sandbag was released. (g = 9·8 m s^{-2}.)

Worked example

34 *A body is projected from the top of a tower 24·5 m high with a velocity of 39·2 m s^{-1} at an angle of 30° with the horizontal. Find (i) the time taken for it to strike the ground, (ii) the distance from the foot of the tower when the impact occurs, (iii) the velocity of the body on impact with the ground, (iv) the direction in which the body is then travelling, (v) the greatest height above the ground attained by the body.*

Let the vertical and horizontal components of the velocity V at the

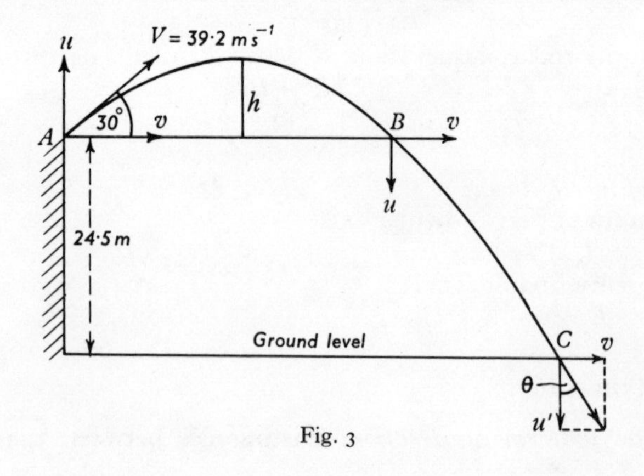

Fig. 3

point of projection (A in the diagram) be respectively

$u\ (= V \sin 30 = 39{\cdot}2 \times \frac{1}{2} = 19{\cdot}6 \text{ m s}^{-1})$

and $v\ (= V \cos 30 = 39{\cdot}2 \times \frac{\sqrt{3}}{2} = 19{\cdot}6\sqrt{3} \text{ m s}^{-1})$

At the point B on the horizontal plane through A the vertical component of the body's velocity is u, but now in a downward direction; whilst at the point C, where the body strikes the ground, the vertical component of the velocity is u' (say). The horizontal component of velocity is not subject to acceleration and has, therefore, the same value v throughout the motion.

(i) Let t = time taken for the body to travel from A to C. During this time the body travels a vertical distance of 24·5 m (since C is 24·5 m vertically below the zero at A). Hence (applying equations of linear motion)

$$-24{\cdot}5 = 19{\cdot}6\,t - \tfrac{1}{2} \times 9{\cdot}8\,t^2$$

$$\text{i.e.} \qquad t^2 - 4t - 5 = 0$$

from which $t = \underline{5}$ s (ignoring the negative time value).

(ii) During the above time the horizontal velocity remains constant at $19{\cdot}6\sqrt{3}$ m s^{-1}. Hence distance of C from foot of tower

$$= 19{\cdot}6\sqrt{3} \times 5 = \underline{169{\cdot}7 \text{ m}}$$

(iii) Let the velocity of the body on impact at C be V', then $V' = \sqrt{u'^2 + v^2}$ where $v = 19{\cdot}6\sqrt{3}$ m s^{-1}. During the flight of 5 seconds the upward velocity of 19·6 m s^{-1} at A is retarded by gravity at the rate of 9·8 m s^{-1} per second. Hence

$$u' = 19{\cdot}6 - 9{\cdot}8 \times 5 = -29{\cdot}4 \text{ m s}^{-1}$$
$$= 29{\cdot}4 \text{ m s}^{-1} \text{ in a } \textit{downward} \text{ direction}$$

and accordingly

$$V' = \sqrt{(29{\cdot}4)^2 + (19{\cdot}6\sqrt{3})^2}$$
$$= \underline{44{\cdot}9 \text{ m s}^{-1}}$$

(iv) Let the body be travelling in such a direction at C that it makes an angle of θ with the vertical. Then

$$\theta = \tan^{-1}\frac{v}{u'} = \tan^{-1}\frac{19{\cdot}6\sqrt{3}}{29{\cdot}4} = \tan^{-1}\frac{2\sqrt{3}}{3}$$
$$= \underline{49°6'}$$

(v) The greatest height above the ground is clearly $(24{\cdot}5 + h)$ m where h = greatest height of the parabolic trajectory A to B. Using the expression $\frac{V^2 \sin^2\theta}{2\,g}$ for the greatest height reached in the parabolic path we have

$$h = \frac{V^2 \sin^2 30}{2\,g} = \frac{39{\cdot}2 \times 39{\cdot}2 \times \frac{1}{4}}{2 \times 9{\cdot}8} = 19{\cdot}6 \text{ m}$$

Hence, greatest height reached above ground

$$= 24{\cdot}5 + 19{\cdot}6 = \underline{44{\cdot}1} \text{ m}$$

35 Prove that, neglecting air resistance, the path of a heavy projectile is a parabola.

A stone is projected at an angle of 60° to the horizontal with a velocity of 50 m s^{-1}.

Calculate (*a*) the highest point reached, (*b*) the range, (*c*) the time taken for the flight, (*d*) the height of the stone at the instant its path makes an angle of 30° with the horizontal. (g = 9·8 m s^{-2})

36 Show that the velocity of a projectile at any point on its trajectory is equal to that which would be acquired by a particle falling freely from the height of the directrix of the parabolic trajectory at the point in question.

An object is projected so that it just clears two obstacles, each 25 m high, which are situated 160 m from each other. If the time of passing between the obstacles is 2·5 s, calculate the full range of projection and the initial velocity of the object. (g = 9·8 m s^{-1})

37 Show that the greatest range of a body projected on an inclined plane is obtained when the angle of projection bisects the angle between the vertical and the inclined plane and show further that this range is equal to the distance through which the body would fall freely during the corresponding time of projection.

Compare the ranges on an inclined plane making an angle of 30° with the horizontal when a body is projected at an angle of 60° with the horizontal (*a*) up the plane, (*b*) down the plane, the velocity of projection being the same in each case.

38 An explosive device is so designed that when it is detonated it fragments into a number of pieces of equal mass. Assuming the energy of the explosion is equally divided amongst these fragments, find the maximum spread between them when the device is exploded at ground level and a vertically flying fragment is observed to attain a height of 25 metres. Ignore the effect of air resistance.

39 State Newton's laws of motion and give an example in illustration of each law.

What force is needed to drive a lorry of mass 5000 kg up a slope of 1 in 20 with an acceleration of 0·5 m s^{-2} if opposing forces due to air resistance and road friction amount to 75 N per 1000 kg of lorry mass? (g = 9·80 m s^{-2})

40 A vehicle of mass 1000 kg travelling along a horizontal road surface at 40 km hr^{-1} suddenly meets an incline of 1 in 10. If the vehicle's engine is shut off immediately it meets the incline, how far can it travel up it before coming to rest? Ignore all frictional and other forces opposing the motion.

41 A car of mass 1500 kg coasts, from rest, down a hill of gradient 1 in 8. If the velocity of the car after travelling 400 metres down the hill is 20 m s^{-1} estimate an average value of the forces opposing the car's motion.

Worked example

42 *A mass m_1 of 200 g is placed on a smooth plane inclined at 30° with the horizontal. To this mass is attached a string which, after passing over a small smooth pulley at the top of the plane, freely supports another mass m_2 of 105 g hanging vertically. Calculate the distance m_2 descends from rest in 2 s and the tension in the string. (g = 9·80 m s^{-2})*

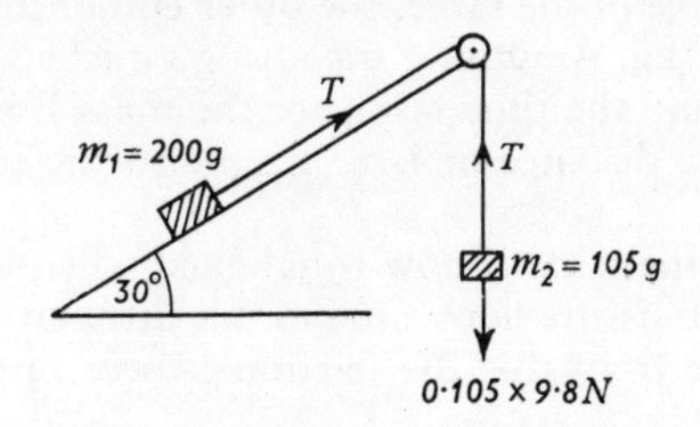

Fig. 4

Let the tension in the string be T and let a be the common acceleration of m_1 up the plane and m_2 vertically. Then, for the motion of m_2 we have

$$0{\cdot}105 \times 9{\cdot}8 - T = 0{\cdot}105\, a \qquad (1)$$

and for the motion of m_1 (for which the resolved part of the weight down the plane is $0{\cdot}200 \times 9{\cdot}8 \times \sin 30$) we have

$$T - 0{\cdot}200 \times 9{\cdot}8 \times \sin 30 = 0{\cdot}200\,a \qquad (2)$$

Adding (1) and (2) we get

$$(0{\cdot}105 - 0{\cdot}200 \sin 30)\,9{\cdot}8 = 0{\cdot}305\,a$$

from which $$a = \frac{0{\cdot}005 \times 9{\cdot}8}{0{\cdot}305} = \underline{0{\cdot}16}\ \text{m s}^{-2}$$

Hence, distance m_2 descends in 2 seconds

$$= \tfrac{1}{2}at^2 = \tfrac{1}{2} \times 0{\cdot}16 \times 2^2 = \underline{0{\cdot}32}\ \text{m}$$

From (1) we have

$$T = 0{\cdot}105\,(9{\cdot}8 - a)$$
$$= 0{\cdot}105\,(9{\cdot}8 - 0{\cdot}16) = \underline{1{\cdot}01}\ \text{N}$$

43 Describe critically the method of determining the acceleration of gravity using Atwood's machine.

In a simple Atwood's machine masses of 0·5 kg and 0·48 kg are hung over a frictionless pulley of negligible mass. Calculate the tension in the string and the displacement of the centre of gravity of the system 3 seconds after setting the system in motion. (Take g as 9·8 m s^{-2})

44 A block A of mass 0·5 kg rests on a smooth horizontal table. A light inextensible string attached to A passes over a frictionless pulley mounted at the edge of the table, the other end of the string being tied to a mass B of 0·1 kg. Assuming the string attached to A runs parallel to the table top, find the time taken for the mass B to fall from its rest position through a distance of 1 m. Find also the tension in the connecting string.

The surface of the table is now roughened when it is found that the mass B takes twice its previous time to fall through the measured 1-m distance. Estimate from this the frictional force opposing the motion of A.

45 Define *momentum*. How is it related to force?

Water issuing at the rate of 5 m s^{-1} from a pipe 10 cm in diameter is directed onto a metal plate situated close to the efflux end of the pipe. Assuming the water stream to strike the plate normally, calculate the thrust sustained by the plate. (Density of water = 10^3 kg m^{-3})

46 State the law of *conservation of momentum* and describe a laboratory method of verifying the law.

A 100-kg hammer falls from a vertical height of 5 m from rest and drives a stake into the ground in a time interval of 0·10 s from initial impact. What is the average force of resistance of the ground? ($g = 9{\cdot}8$ m s^{-2})

47 Define the term *power* as used in mechanics.

Find the power required to propel a motor-car at a steady speed of 30 kilometres per hour if, at that speed, the force resisting the motion is equal to the weight of 200 N.

48 Define *momentum* and *kinetic energy*.

A body of mass 50 g falling freely from rest from a position 30 cm vertically above a horizontal surface, rebounds to a height of 20 cm after impact. Calculate the change in momentum and kinetic energy on impact. ($g = 9{\cdot}8$ m s^{-2})

49 A bullet of mass 20 g, moving at 50 m s^{-1} embeds itself in a fixed target to a depth of 2·5 cm. Calculate (*a*) the kinetic energy of the bullet immediately before entering the target, (*b*) the average resisting force experienced on entering the target.

50 What do you understand by (*a*) the *principle of conservation of energy*, (*b*) *the principle of conservation of momentum?*

A bullet of mass 20 g is fired horizontally into a 1-kg block of wood suspended by metre-long light vertical strings. If, after the bullet has embedded itself into the block, the strings are deflected through an angle of 30° with the vertical, calculate the velocity of the bullet just before impact with the block.

To what extent do the above two physical principles apply to this problem?

Worked example

51 *State clearly the two conditions which determine the velocities of two freely impacting elastic bodies. Apply these conditions to find the velocities after collision of two steel spheres, one of which has a mass of* 0·5 *kg and is moving with a velocity of* 0·5 *m s*$^{-1}$ *to overtake the other sphere which has a mass of* 0·2 *kg and is moving in the same direction as the first sphere with a velocity of* 0·3 *m s*$^{-1}$. *The coefficient of restitution is* 0·75.

The conditions determining the velocities of the colliding spheres are:

(1) *Newton's law of restitution,* viz. the relative velocity of the spheres

after collision = $-e\times$ the relative velocity of the spheres before collision (e being the coefficient of restitution),

i.e. $$v_1 - v_2 = -e\,(u_1 - u_2)$$

or, inserting numerical values,

$$v_1 - v_2 = -0{\cdot}75\,(0{\cdot}5 - 0{\cdot}3) = -0{\cdot}15 \qquad \text{(i)}$$

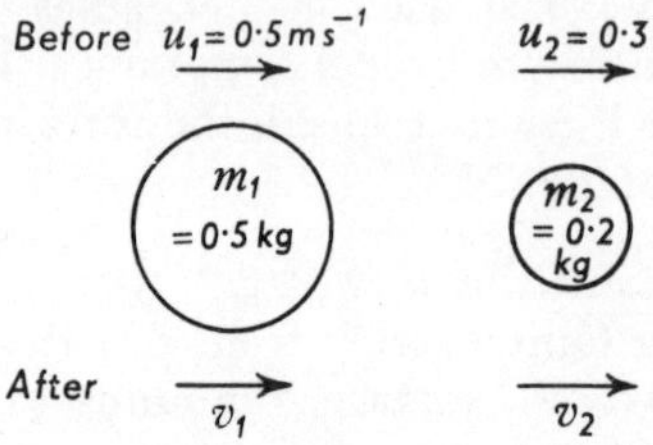

Fig. 5

(2) *The law of conservation of momentum*—the total momentum of the system after impact is equal to the total momentum of the system before impact.

i.e. $$m_1 v_1 + m_2 v_2 = m_1 u_1 + m_2 u_2$$

or, inserting numerical values,

$$\begin{aligned} 0{\cdot}5\,v_1 + 0{\cdot}2\,v_2 &= 0{\cdot}5\,u_1 + 0{\cdot}2\,u_2 \\ &= 0{\cdot}5 \times 0{\cdot}5 + 0{\cdot}2 \times 0{\cdot}3 \\ &= 0{\cdot}31 \qquad \text{(ii)} \end{aligned}$$

Now multiplying equation (i) by $0{\cdot}2$ we have

$$0{\cdot}2\,v_1 - 0{\cdot}2\,v_2 = -0{\cdot}03$$

And adding this equation to equation (ii) to eliminate v_2 we get

$$0{\cdot}7\,v_1 = 0{\cdot}28$$

giving $$v_1 = \underline{0{\cdot}4}\ \mathrm{m\,s^{-1}}$$

Inserting this value for v_1 in equation (i) we get

$$v_2 = \underline{0{\cdot}55}\ \mathrm{m\,s^{-1}}$$

these being the velocities of the $0{\cdot}5$ kg and $0{\cdot}2$ kg spheres respectively after collision, both velocities being in the same direction as the original velocities.

52 Prove that when two imperfectly elastic bodies collide, a loss of energy occurs.

Calculate the rise in temperature of the spheres in the above problem if the specific heat capacity of steel is 462 J kg^{-1} K^{-1}.

53 What do you understand by the *coefficient of restitution?* How would you find the value of this coefficient for two glass spheres?

A glass marble falls from a height of 1 metre on to a hard surface and the height of the second rebound is found to be 45 cm. What is the value of the coefficient of restitution for the marble and the surface?

54 State the conditions which determine the after collision velocities of two elastic bodies undergoing linear impact.

A helium atom travelling with a velocity of U impacts directly on a stationary hydrogen atom. On the assumption that the collision is perfectly elastic, and that the helium atom has exactly four times the mass of the hydrogen atom, calculate (*a*) the percentage change in the energy of the helium atom and, (*b*) the velocity of the hydrogen atom as a result of the collision.

55 Two impact trolleys, A and B, have masses m, M respectively ($M > m$). B is at rest and A is directed towards it with a velocity of u. If the coefficient of restitution for the trolleys is e, obtain an expression for the velocity of A after impact in terms of the quantities involved. Hence, or otherwise, find
(*a*) the ratio of the impacting masses for a system for which $e = 0{\cdot}5$ if A is brought to rest after colliding with B,
(*b*) the value of e if A rebounds from B with half its initial velocity, the ratio of the mass of B to that of A being, in this case, 3 : 1.

56 A steel sphere falls from a height of 2 metres on to a hard surface for which the coefficient of restitution is 0·8. Calculate the total time for which the motion ensues and the total distance covered by the bouncing sphere in this time. ($g = 9{\cdot}8$ m s^{-2})

57 Two smooth spheres moving in opposite parallel direction with equal velocities of 20 cm s^{-1} collide in such a way that their directions of motion at the moment of impact make angles of 30° with the line of their centres. If the mass of one sphere is double that of the other, and the coefficient of restitution between them is 0·5, find the velocities and direction of motion of the spheres after impact.

58 From a point on a smooth plane a particle is projected with a velocity of 20 m s^{-1} at an angle of 30° with the horizontal. If the coefficient of restitution between the particle and the plane is 0·5, find the time elapsing before the particle ceases to rebound and the distance described by the particle along the plane during this time.

Uniform circular motion

59 Derive an expression for the acceleration of a body moving along a circular path with uniform speed.

The bob of a metre-long simple pendulum has a mass of 20 g and is made to move round a horizontal circular path of radius 10 cm. Calculate (*a*) the speed of the bob, (*b*) the tension in the string.

60 What is the 'angle of banking'? How is it related to the speed of the vehicle and the radius of the track along which it is travelling?

Calculate the angle of banking for a railway track following a curved course of radius 400 m if the specified speed along the track is to be 80 km hr^{-1}.

61 What do you understand by the term 'centripetal force'? Give **two** examples to illustrate your statement.

Calculate the radius of the sharpest curve which can be negotiated without skidding by a vehicle travelling at a speed of 60 km hr^{-1} on a level surface if the coefficient of friction between the wheels and the surface is 0·5. ($g = 9{\cdot}8$ m s^{-2})

62 Calculate the maximum number of revolutions per minute at which a body of mass 200 g, attached to the end of a string of length 1 metre, can be whirled in a horizontal plane if the breaking tension in the string is 49 newtons. In what direction will the body travel when the string breaks?

Worked example

63 *The semi-vertical angle of a conical pendulum of length* 1·5 *m is* 30°. *What is its period of revolution? If the mass of the bob is* 0·2 *kg, calculate the tension in the string when the pendulum is revolving as above.*

The diagram shows the motion of the bob and the forces acting on it. If the speed of the bob on its circular path of radius r is v, then the time of revolution $= \dfrac{2\pi r}{v}$. Now resolving the forces at B vertically and horizontally, we have, vertical component of tension = force of gravity on bob

i.e. $$T \cos \theta = mg \qquad \text{(i)}$$

horizontal component of tension = centripetal force on bob

i.e. $$T \sin\theta = \frac{mv^2}{r} \qquad \text{(ii)}$$

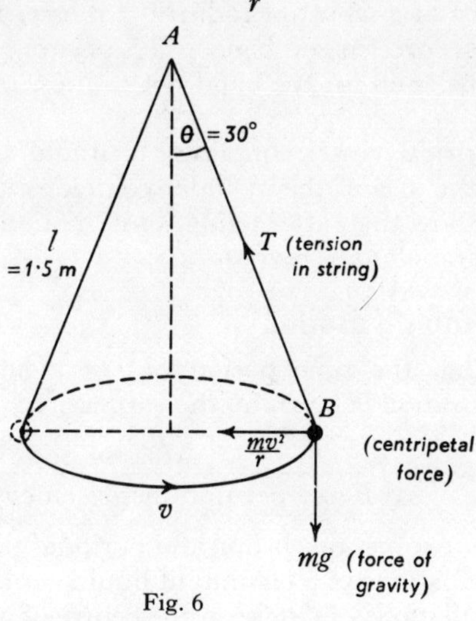

Fig. 6

Dividing equation (ii) by equation (i), we get

$$\tan\theta = \frac{v^2}{rg}$$

or $$v = \sqrt{rg\tan\theta}$$

Hence the time of revolution may be written

$$\frac{2\pi r}{\sqrt{rg\tan\theta}} = 2\pi\sqrt{\frac{r}{g\tan\theta}} = 2\pi\sqrt{\frac{l\sin\theta}{g\tan\theta}} = 2\pi\sqrt{\frac{l\cos\theta}{g}}$$

Inserting numerical values we thus get for the time of revolution

$$2\pi\sqrt{\frac{1{\cdot}5\times\cos 30}{9{\cdot}8}} = \underline{2{\cdot}26\ \text{s}}$$

From equation (i) the tension T in the string is

$$\frac{mg}{\cos\theta} = \frac{0{\cdot}2\times 9{\cdot}8}{\cos 30} = \underline{2{\cdot}26\ \text{N}}$$

64 What is the least velocity a trick cyclist needs on entering the loop of a 'loop-the-loop' track of radius 4 m if he is to successfully complete the manoeuvre?

65 Discuss the concept of 'centrifugal force' illustrating your answer by two cases in which you consider it may usefully be applied.

An elastic band has a mass 0·1 kilogramme and is stretched on the circumference of a wheel of radius 0·2 metre. If, when so positioned, the stretching force in the band is 2·5 newton, find at what speed the wheel must be spun for the band not to press on it.

66 A cylindrical vessel containing a liquid is placed on a whirling disc so that the axis of the cylinder coincides with the axis of rotation of the disc. Show that, after being set in motion, the free surface of the liquid is a paraboloid of revolution.

Simple harmonic motion

67 Show that the time period (T) of a body performing simple harmonic vibration is given by the expression

$$T = 2\pi \sqrt{\frac{\text{Mass of body}}{\text{Force per unit displacement of the body}}}$$

Apply this expression to find the periodic time of the following:
(a) the oscillations of a column of liquid contained in a U-tube,
(b) the oscillations of a mass at the centre of a tight wire.

68 Explain what you understand by *simple harmonic motion.*

Show that if a point moves with uniform speed in a circle, its projection on a diameter of that circle moves with simple harmonic motion.

A body executes linear simple harmonic vibration about a certain point. If the velocity and acceleration of the body when displaced 5 cm from this point are respectively 17·3 cm s^{-1} and 20 cm s^{-2}, find the amplitude and periodic time of the motion.

Worked example

69 *A particle executing simple harmonic vibrations in a straight line has velocities of* 8 *cm* s^{-1} *and* 2 *cm* s^{-1} *when at positions* 3 *cm and* 5 *cm respectively from its equilibrium position. Calculate the amplitude of the vibrations and the periodic time of the particle.*

Let the point P move with uniform circular motion in a circle of radius a (see diagram), then the projection of this motion on a diameter will be simple harmonic. Thus the point N executes simple harmonic vibration of amplitude a about the point O. The velocity of N at a displacement x from O is the horizontal component of the velocity of P at the corresponding point. Thus

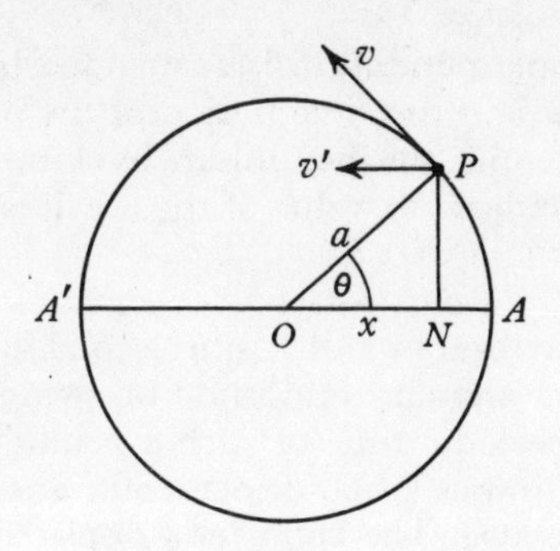

Fig. 7

$$v' = v \sin\theta = \frac{v}{a}\sqrt{a^2 - x^2}$$

Hence, using the data supplied,

$$8 = \frac{v}{a}\sqrt{a^2 - 9} \qquad (1)$$

and

$$2 = \frac{v}{a}\sqrt{a^2 - 25} \qquad (2)$$

Squaring and dividing these equations we get

$$\frac{64}{4} = \frac{a^2 - 9}{a^2 - 25}$$

from which $a^2 = \dfrac{391}{15}$ giving $a = \underline{5{\cdot}1}$ cm

The periodic time (T) of the simple harmonic vibration will be equal to the time taken by P to completely travel round the circle.

Thus $$T = \frac{2\pi a}{v}$$

Now from equation (1) above, $\dfrac{a}{v} = \dfrac{\sqrt{a^2 - 9}}{8} = \dfrac{\sqrt{(5{\cdot}1)^2 - 9}}{8}$

Hence $T = \dfrac{2\pi\sqrt{(5{\cdot}1)^2 - 9}}{8} = \underline{3{\cdot}24}$ s

70 Critically discuss the errors involved in a determination of the acceleration of gravity from observations of the time of swing of a simple pendulum.

The bob of a simple pendulum is suspended by a long string from an inaccessible point and the time period of the pendulum is found to be 4·50 s. On shortening the pendulum by 1 metre, the time period becomes 4·03 s. Calculate the value of the acceleration of gravity at the place concerned.

71 Show that the vibrations of a simple pendulum are simple harmonic providing the angular amplitude of swing is small. Derive an expression for the periodic time of such a pendulum.

A simple pendulum has a bob of mass 50 g suspended at the end of a light string 1 metre long. The bob is now displaced so that the taut string makes an angle of 10° with the vertical. On release of the bob calculate (i) the time period of the pendulum, (ii) the velocity of the bob as it passes through the centre point of the oscillation, (iii) the tension in the string at this latter point.

72 A volume V of air is contained at atmospheric pressure P in a cylindrical vessel of cross-sectional area A by a frictionless air-tight piston of mass M. Show that, on slightly forcing down the piston and then releasing it, the piston oscillates with simple harmonic motion of period $\frac{2\pi}{A}\sqrt{\frac{MV}{P}}$

(Assume the conditions to be isothermal.)

73 A test-tube is loaded with lead shot so that it floats vertically in a liquid. Show that after being further immersed in the liquid, the vertical oscillations of the tube that result after removing the immersing force are simple harmonic.

The periodic time for vertical oscillations of such a tube when immersed in water is found to be 1·20 s, and when immersed in paraffin is 1·34 s. What is the relative density of paraffin?

74 Show that the total energy of a body executing simple harmonic motion is constant and proportional to the square of the amplitude.

A particle of mass 2 g executes linear simple harmonic vibrations of periodic time 2 s and amplitude 10 cm. Find the kinetic energy of the particle when passing through its equilibrium aposition and when displaced 5 cm from that position.

75 If the free end of a vibrating cantilever has a maximum vertical movement of 20 cm, calculate its shortest permissible period of vibration if an object resting on the end of the cantilever is to remain in contact with it throughout the motion.

76 Show that the vertical oscillations of a loaded helical spring are simple harmonic, and describe how a value for the acceleration of gravity may be obtained from observations of the time period of such a loaded spring.

A spiral spring extends 5 cm when a load of 100 g is attached. Calculate the maximum extension of this spring if a mass of 50 g is dropped from a height of 10 cm on to a light pan attached to the spring.

77 A cylindrical vessel with an area of cross-section A contains a volume V of gas at a pressure p which is just sufficient to support a piston of mass M which slides freely in the cylinder. The piston is given a slight displacement and subsequently released. Show that its motion will then be simple harmonic and obtain an expression for the periodic time of this motion if the pressure-volume relation of the gas during the accompanying changes is given by the expression $pV^{\gamma} = \text{const.}$

Find the periodic time of the motion if $M = 0{\cdot}1$ kg, $A = 10^{-3}$ m^2, $V = 10^{-3}$ m^3, $\gamma = 1{\cdot}4$ and the external pressure is equivalent to 0·76 m of mercury. (Take the value of the acceleration of gravity as 9·8 m s^{-2} and the density of mercury as $13{\cdot}6 \times 10^3$ kg m^{-3}.)

78 A simple pendulum is set swinging between the 10 and the 30 division markings of a horizontal scale clamped behind the pendulum. A photograph taken of the moving pendulum shows a trace of the thread extending between the 18 and the 25 division markings on the scale. If the pendulum has a period of 1·50 s, calculate the shutter speed of the camera.

79 What do you understand by *compound harmonic motion, Lissajous' figures?* Describe how you could obtain Lissajous' figures in the laboratory, and state what information can be obtained from such experimental patterns.

A particle is subject to two simple harmonic vibrations applied simultaneously along directions mutually at right angles to each other. Discuss the resultant track of the particle if the two motions have a frequency ratio of 2:1 (*a*) when they are in phase, (*b*) when there is a phase difference of π. Discuss also the effect of differing amplitudes in the two motions.

80 A sphere of radius r rolls without slipping on a concave surface of large radius of curvature R. Show that the motion of the centre of gravity of the sphere is approximately simple harmonic with a time period of $2\pi\sqrt{\dfrac{7(R-r)}{5g}}$

Rotational dynamics

81 Obtain an expression for the kinetic energy of a rotating body.

A uniform disc of mass 100 g has a diameter of 10 cm. Calculate the total energy of the disc when rolling along a table with a velocity of 0·2 m s^{-1}.

82 Explain the term *radius of gyration*.

The axle of a wheel and axle is supported on a pair of parallel rails which are inclined at an angle of 10° with the horizontal. If the diameter of the axle is 1 cm, and the wheel and axle rolls from rest (without slipping) a distance of 100 cm down the rails in 8 s, what is its radius of gyration?

83 Obtain an expression for the acceleration of a body rolling down a plane inclined at an angle of θ with the horizontal.

A hoop, a disc and a solid sphere, all of the same diameter, are set rolling down an inclined plane which is sufficiently rough to prevent slipping. Compare the times taken for each body to roll (from rest) down the plane with that taken by a body to slide (also from rest) down a smooth plane of the same length and slope.

84 What do you understand by the term 'moment of inertia' of a body? Establish the theorem of 'perpendicular axes' relating to moments of inertia through the mass centre of a body.

A thin, uniform metal disc is (*a*) spun about a diametral axis, (*b*) rotated about an axis through its mass centre perpendicular to its plane, (*c*) rolled along a horizontal surface. Compare the kinetic energy of the disc in each of the three cases if its rotational velocity is constant throughout.

85 A light inextensible string passes round the edge of a flat circular disc pulley of mass 25 g and radius 5 cm. The ends A and B of the string support equal masses each of 100 g, and when an additional mass of 1 g is applied at B, movement of the system ensues resulting in the mass attached at B falling a vertical distance of 2 m in 6·55 s. From these observations calculate a value for the acceleration of gravity (*a*) ignoring the effect of the pulley disc, (*b*) taking it into account.

Worked example

86 *A flywheel with an axle* 2·0 *cm in diameter is mounted on frictionless bearings. A light inextensible cord is wrapped round the axle and supports a mass of* 10g. *On being released, the mass falls through a distance of* 2 *m* *in 10 s after which it becomes detached. Find (a) the torque producing the motion of the flywheel whilst the weight is falling, (b) the moment of inertia of the flywheel, (c) the kinetic energy of the flywheel at the moment when the attached mass is detached, (d) the constant retarding torque which would bring the flywheel to rest in 1 revolution after the thread and attached mass have been detached.*

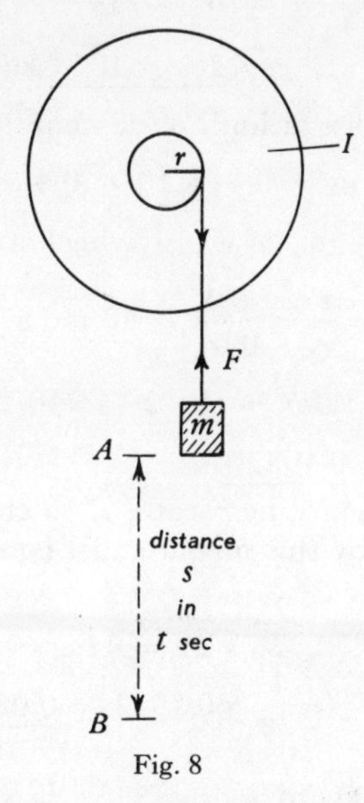

Fig. 8

(*a*) Let the tension in the string be F, then for the motion of the attached mass we have

$$mg - F = ma$$

Now the acceleration a of the mass is given by $s = \frac{1}{2} at^2$

so
$$a = \frac{2s}{t^2}$$

and hence

$$F = m(g-a) = m\left(g - \frac{2s}{t^2}\right)$$

$$= 0{\cdot}01\left(9{\cdot}84 - \frac{2\times 2}{10\times 10}\right)$$

$$= 0{\cdot}0976 \text{ N}$$

Accordingly, the propelling torque $T(= F\times r)$

$$= 0{\cdot}0976\times 0{\cdot}01 = \underline{0{\cdot}000976} \text{ N m}.$$

(*b*) Now for the motion of the flywheel $T = I\alpha$ where α is the angular acceleration of the wheel which is

$$\frac{a}{r} = \frac{0{\cdot}04}{0{\cdot}01} = 4 \text{ rad s}^{-2}$$

$$\text{Hence} \quad I = \frac{T}{\alpha} = \frac{0{\cdot}000976}{4} = 0{\cdot}000244$$

$$= \underline{2{\cdot}44 \times 10^{-4}} \text{ kg m}^2$$

(*c*) Velocity (v) of mass at limit of descent

$$= at = 0{\cdot}04 \times 10 = 0{\cdot}4 \text{ m s}^{-1}$$

Hence angular velocity (ω) of the flywheel at this point

$$= \frac{v}{r} = \frac{0{\cdot}4}{0{\cdot}01} = 40 \text{ rad s}^{-1}$$

$$\text{So K.E. of flywheel} = \tfrac{1}{2}I\omega^2 = \tfrac{1}{2} \times 2{\cdot}44 \times 10^{-4} \times 40^2$$

$$= \underline{0{\cdot}1952 \text{ J}}$$

(*d*) For a constant retarding torque T' to check the wheel in 1 revolution the work done by this torque must equal the K.E. of the wheel at (*c*). That is,

$$T' \times 2\pi = \tfrac{1}{2}I\omega^2$$

so

$$T' = \frac{1}{2\pi} \times 0{\cdot}1952 = \underline{0{\cdot}0311} \text{ N m}$$

87 A uniform cylinder of radius 5 cm is set to roll down a plane inclined at an angle of 30° with the horizontal. If slipping just does not occur, find the coefficient of friction between the cylinder and the plane and the angular acceleration of the cylinder.

88 A solid cylindrical object (*a*) slides, without rolling, down a smooth plane, (*b*) rolls, without sliding, after the surface of the plane has been suitably roughened. If, in each case, the cylinder starts from rest, and the angle of inclination of the plane is 30° with the horizontal, find the time taken to travel 2 metres down the plane in the two cases. ($g = 9{\cdot}80 \text{ m s}^{-2}$)

89 Describe, with full experimental details, how you would determine the moment of inertia of a fly wheel.

A flywheel of moment of inertia 10^{-2} kg m^2 rotating at 50 rev. s^{-1} is brought to rest by the friction of the bearings after completing a further 100 revolutions. Calculate the frictional couple exerted by the bearings and, assuming the value of this couple to be constant, find the horse-power needed to keep the wheel rotating at 100 rev. s^{-1}. (1 h.p. = 746 watt)

90 A flywheel in the form of a uniform solid disc is mounted on a light axle of radius 2 cm round which is wound a cord to which is attached a mass of 0·5 kg. If the thickness of the flywheel is 4 cm, its radius 10 cm and its density 7·8 g cm^{-3}, find the tension in the cord and the kinetic energy of the flywheel when the attached mass has descended a distance of 20 cm from rest.

91 A flywheel, of radius 10 cm, is pivoted to run freely on a horizontal axis. A small piece of wax, of mass 1 g, is attached to the edge of the wheel which, after a slight displacement, is found to execute oscillations of time period 10 s. Show that the oscillations of the wheel are simple harmonic, and calculate a value for the moment of inertia of the wheel.

92 A metal plate is firmly suspended in a horizontal plane by a torsion wire attached to the plate at its mass centre. The plate is then allowed to make torsional oscillations of small amplitude the observed time period being 4·0 s. A thin metal rim of mass 0·1 kg and radius 5 cm is then placed on the plate so that the centre of the rim is immediately above the centre of oscillation of the plate. The time period of the combination is then found to be 4·5 s. What is the moment of inertia of the plate about the axis of oscillation?

93 A flywheel of moment of inertia 0·1 kg m^2 is set in motion with an angular velocity of 10 rev s^{-1}. It is then left to itself when it is found that after a lapse of 30 s its speed falls to 5 rev s^{-1}. Calculate:
(*a*) the initial kinetic energy of the wheel,
(*b*) the constant retarding torque opposing the motion of the wheel,
(*c*) the total time taken for the wheel finally to come to rest.

94 A wooden cylinder of radius 2 cm has a light inextensible string wrapped round its circumference near its centre. The free end of the string is held firmly in the hand and the cylinder allowed to unwind as it falls to the ground 2 metres below. Assuming there is no slip between the string and the surface of the cylinder, and that the axis of the cylinder remains horizontal as it unwinds, find the time taken for the descent. (g = 9·80 m s^{-2})

95 A mass of 20 g is attached to a length of fine cord which is wrapped round the 3·5 cm diameter axle of a mounted flywheel. The length of the cord is adjusted so that when the attached mass reaches the ground the cord detaches itself from the axle. The distance the mass descends from rest is 2·2 m and it is observed that the descent takes 11 s. It is also observed that the flywheel makes 40 complete revolutions after the mass is detached before the wheel comes completely to rest.

Calculate a value for the moment of inertia of the flywheel from these observations.

96 Given the moment of inertia of a uniform lamina about an axis through its centre of gravity, show how to obtain its moment of inertia (*a*) about a parallel axis in the plane of the lamina, (*b*) about a perpendicular axis through its centre of gravity.

Compare the times of oscillation of a uniform disc when oscillating about an axis on its circumference (i) parallel, (ii) perpendicular, to the plane of the disc.

97 Explain what you understand by the *equivalent simple pendulum* of a rigid or compound pendulum.

Calculate the length of the equivalent simple pendulum for a pendulum consisting of a solid sphere of radius 10 cm suspended by a light wire of length 1 m.

98 Obtain an expression for the time period of a rigid pendulum oscillating about an axis distant h from its mass centre.

What would be the periodic time of a thin metal disc of radius 10 cm oscillating about an axis perpendicular to its plane and passing through a small hole near its periphery?

Worked example

99 *A thin uniform rod is pivoted about a horizontal axis which passes through a point on the rod* 20 *cm from its centre of gravity. If the time of small oscillations performed by the rod in the vertical plane through the suspension is* 1·37 *s, calculate the length of the rod.* ($g = 9{\cdot}81$ $m\ s^{-2}$)

The periodic time of a bar pendulum is given by

$$T = 2\pi\sqrt{\frac{h^2+k^2}{gh}}$$

where h is the distance of the axis of suspension from the C.G. of the bar and k is the radius of gyration of the bar about an axis through its C.G. Hence we have

$$1{\cdot}37 = 2\pi\sqrt{\frac{(0{\cdot}20)^2+k^2}{9{\cdot}81\times 0{\cdot}20}}$$

$$\text{or}\quad k^2 = \frac{(1{\cdot}37)^2\times 9{\cdot}81\times 0{\cdot}20}{4\pi^2} - (0{\cdot}20)^2$$

$$\text{from which}\quad k = \underline{0{\cdot}231\text{ m}}$$

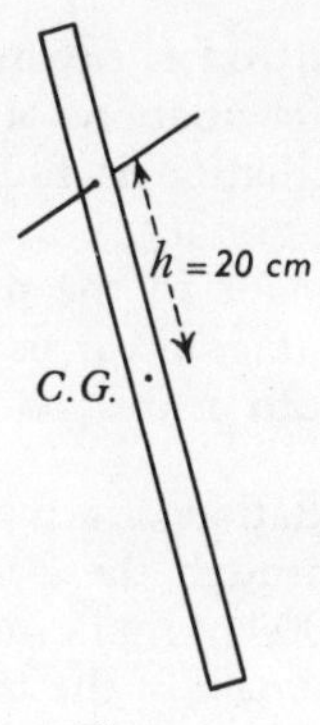

Fig. 9

Now, for a uniform rod of length l the radius of gyration $k = \dfrac{l}{\sqrt{12}}$

$$\text{Hence length of rod} = 0{\cdot}231\times\sqrt{12}$$
$$= \underline{0{\cdot}80\text{ m}}$$

100 A uniform rod is provided with a moveable pivot so that the time of oscillation of the rod can be taken about axes at different positions along the rod. The minimum time period of oscillation for the rod is found to be 1·85 s. What is the length of the rod?

101 Derive expressions for the moment of inertia of (*a*) a disc, (*b*) a rod, about axes through the mass centres perpendicular to their main sections.

A compound pendulum is contrived from a metre rule of breadth 3 cm and mass 24 g by having a circular disc of mass 20 g and with radius 5 cm attached firmly at its circumference to the 100 cm end so that the disc is coplanar with the rule. If the pendulum is suspended from the 25 cm mark so that it can perform oscillations in the common plane of the disc and rule, calculate the periodic time of these oscillations.

102 Compare the merits of a simple pendulum and a compound (rigid) pendulum for a laboratory determination of a value of the acceleration of gravity.

The time period of a compound pendulum of mass 105 g is 1·56 s about parallel axes 20·4 cm and 40·0 cm measured from the mass centre G on opposite sides of it. Find (*a*) the value of the acceleration of gravity, (*b*) the moment of inertia of the pendulum about a parallel axis through the mass centre.

103 A metre-long uniform rod is pivoted about a knife edge 20 cm from one end and performs oscillations of small amplitude in a vertical plane. If the period of oscillation is found to be 1·52 s, calculate a value for the acceleration of gravity.

If an error of 1 mm is made in the positioning of the knife edge, what is the resulting percentage error in the calculated value of g?

(You are to ignore the width of the rod in making your calculations.)

104 Find the time of oscillation of a metre rule when pivoted about a horizontal axis passing through the 20-cm mark. If a mass of 20 g attached to the rule at the 100-cm mark increases the time of oscillation by 10 per cent calculate the mass of the rule.

105 Give the details of a method of finding the acceleration of gravity (g) using a compound pendulum.

Show that, for a reversible pendulum, the value of g is given to a very close approximation by the expression

$$g = \frac{8\pi^2 l}{T_1^2 + T_2^2}$$

where l is the distance between the two knife edges which have been adjusted so that the respective times of swing (T_1 and T_2) about them are very nearly equal.

Under what conditions would the above expression not give a satisfactory value for g?

106 Ball bearings, rolling over a concave spherical surface, just begin to slip when at a point where the tangent to the spherical surface makes an angle of 30° with the horizontal. What is the coefficient of friction between the bearings and the surface?

107 A solid sphere is projected so that it slides with an initial velocity of 2·5 m s^{-1} over a horizontal surface for which the coefficient of friction is 0·1. Find the time interval elapsing after projection before the sphere rolls without slipping, and also the velocity of the sphere when this condition is attained.

PROPERTIES OF MATTER

Elasticity

1 Define *stress, strain, modulus of elasticity.*

What is Young's modulus of elasticity for a wire of diameter 0·5 mm which is stretched by 0·1 per cent of its initial length by a load of 4 kg?

2 Define *Young's modulus of elasticity* and describe how you would find its value experimentally for a metal available in the form of a wire.

What mass attached to the end of a 2-metre long wire of diameter 1 mm will extend it by 2 mm if Young's modulus for the wire is 7×10^{10} N m^{-2}. (g = 9·8 m s^{-2})

3 Define *yield point, permanent set, elastic limit.* Describe the behaviour of a wire that is gradually loaded to breaking point.

Identical loads are attached to two wires, A and B, which have the same initial length. What is the ratio of their extensions if the diameter of A is three times that of B, and if Young's modulus for A is half that for B?

4 State *Hooke's law* and describe how you would verify it for a length of wire subject *either* to twisting *or* to stretching.

An aluminium wire of cross-section 0·002 cm^2 is firmly attached to a rigid support at its upper end whilst a mass of metal of volume 500 cm^3 is attached at the lower end. When the mass of metal is completely immersed in water the length of the wire is observed to change by 1·0 mm. What is the length of the aluminium wire? (Young's modulus for aluminium = 7×10^{10} N m^{-2})

Worked example

5 *What load attached to the end of a 2-metre length of steel wire of diameter 1 mm will produce an extension of 2 mm if Young's modulus for steel is 2×10^{11} N m^{-2}? Find also the amount of strain energy stored in the loaded wire. (g = 9·8 m s^{-2})*

By definition, Young's modulus

$$= \frac{\text{longitudinal stress}}{\text{longitudinal strain}}$$

$$= \frac{Mg}{\alpha} \Big/ \frac{\delta L}{L} \quad \text{(see diagram)}$$

$$= \frac{Mg}{\pi r^2} \frac{L}{\delta L}$$

i.e. $$2 \times 10^{11} = \frac{M \times 9{\cdot}8}{\pi \times (5 \times 10^{-4})^2} \times \frac{2}{2 \times 10^{-3}}$$

from which $$M = \frac{2 \times 10^{11} \times \pi\,(5 \times 10^{-4})^2 \times 2 \times 10^{-3}}{9{\cdot}8 \times 2}$$

$$= \underline{16{\cdot}03 \text{ kg}}$$

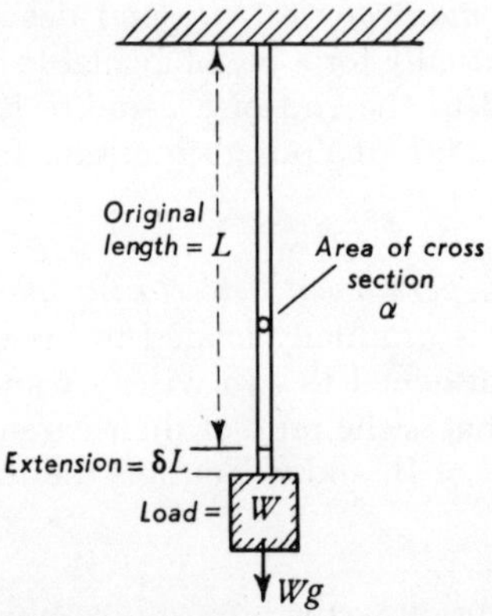

Fig. 10

The energy stored in the wire

$= \frac{1}{2} \times$ final extension $\times$ force producing it

$= \frac{1}{2} \times 2 \times 10^{-3} \times 16{\cdot}03 \times 9{\cdot}8$

$= \underline{0{\cdot}157 \text{ J}}$

6 Obtain an expression for the energy per unit volume of a strained wire in terms of its Young's modulus of elasticity and the strain produced.

A steel wire of length 3 metres and diameter 1 mm is subject to a progressively increasing tensile stress. Calculate the increase in the strain energy stored in the wire as the extension of the wire is increased from 3 mm to 4 mm. (Young's modulus for steel $= 2 \times 10^{11}$ N m^{-2})

7 A wire of diameter 1 mm is held horizontally between two rigid supports 2 metres apart. What mass, attached to the mid-point of the wire, will produce a sag of 5 cm if Young's modulus for the wire is 11×10^{11} N m^{-2}? (Ignore the mass of the wire.)

8 A brass rod, of diameter 5 mm, is heated to a temperature of 300°C when its ends are firmly clamped. Find the force that must be exerted by the clamps on the rod if it is to be prevented from contracting on cooling to 15°C. (Linear expansivity of brass = $0{\cdot}000019°C^{-1}$, Young's modulus for brass = 9×10^{11} N m^{-2}.)

9 Define *bulk modulus.* Derive expressions for the bulk modulus of a perfect gas (i) when the compression takes place under isothermal conditions, (ii) when it takes place under adiabatic conditions.

10 What is *Poisson's ratio*? Describe how you would determine its value experimentally for a substance available in the form of a wire.

A copper wire, 0·5 mm in diameter, extends by 1 mm when loaded with a mass of 0·5 kg and twists through 1 radian when a torque of $5{\cdot}75 \times 10^{-7}$ N-m is applied to its unclamped end. Calculate a value of Poisson's ratio for copper.

11 Define *shearing stress, modulus of rigidity.*

Derive, in terms of the relevant physical quantities, an expression for the torsional couple required to twist a wire through an angle of θ radians.

A 40-cm length of wire of diameter 1 mm is firmly clamped at its upper end whilst its lower end is attached to the centre of a disc of metal of mass 1 kg and radius 7·5 cm. The disc, when displaced, performs torsional oscillations in a horizontal plane, the time period being 2·4 seconds. What is the modulus of rigidity of the wire?

12 Obtain an expression for the sag of a loaded cantilever in terms of the physical quantities involved.

To the free end of a light cantilever in the form of a cylindrical rod of dimensions 40 × 0·5 cm is attached a mass of 200 g. The time for small vertical oscillations of the loaded cantilever is found to be 0·65 s. What is the value of Young's modulus for the cantilever?

13 A metre rule, of breadth 3·0 cm and depth of 4 mm, is supported on knife-edges at the 10 cm and the 90 cm marks. A load of 300 g applied at the mid-point of the rule produces a depression of 1·75 cm. Calculate a value for Young's modulus of the material of the rule. Give the theory of your method.

Surface tension

14 Two clean glass plates are placed vertically parallel to each other with their lower ends dipping into a beaker of water. If the distance between the plates is 0·05 mm, to what height will the water rise between them? (Surface tension of water = 0·072 N m^{-1}, density of water = 10^3 kg m^{-3}, g = 9·80 m s^{-2}.)

15 A glass microscope slide has dimensions $6 \times 2{\cdot}5 \times 0{\cdot}1$ cm and is suspended from one arm of a counterpoised beam balance so that the slide is in a vertical plane with its long side horizontal. A beaker of water placed below the microscope slide is now raised until the water surface just touches the lower edge of the slide when it is noticed that the counterpoise of the balance is upset. Explain why this is, and calculate a value for the surface tension of water if counterpoise is re-established on further raising the beaker so as to submerge $\frac{3}{5}$ths of the slide.

16 Give, with brief explanations, **four** illustrations of the phenomena of surface tension.

A glass capillary tube, having a uniform internal diameter, is placed vertically with one end dipping into paraffin for which the surface tension is 0·027 N m^{-1} and for which the angle of contact is 26° and whose density is 850 kg m^{-3}. If the paraffin rises to a height of 4·5 cm, what is the diameter of the tube?

What happens if the tube is lowered until only the top 3 cm of its length is out of the paraffin?

17 Define *angle of contact* and describe a suitable method for measuring this angle for mercury against glass.

Water rises in a glass capillary tube to a height of 9·6 cm above the outside level. To what depth will mercury be depressed in the same tube if the surface tensions of water and mercury are respectively 0·072 and 0·54 N m^{-1} respectively and their respective angles of contact with glass are 0° and 140°?

Worked example

18 *Two lengths of capillary tubing of diameters* 0·2 *mm and* 1 *mm respectively are joined to make a U-tube in which mercury is placed. What is the difference between the levels of the mercury in the two tubes if the surface tension of mercury is* 0·46 *N* m^{-1} *and its angle of contact with glass* 140°*? (Density of mercury* = $1{\cdot}36 \times 10^4$ *kg* m^{-3}*)*

The excess pressure above atmospheric at a point below a liquid surface convex to the air is given by $\frac{2\gamma}{R}$, where γ is the surface tension of the liquid and R the radius of curvature of the surface. For the convex surface of mercury in a capillary tube in which the angle of contact is 140°, the radius of curvature (R) of the free surface is related to the radius (r) of the capillary tube as follows

$$R = \frac{r}{\cos 40} \quad \text{(see Fig.11(a))}$$

Hence excess pressure at A (see Fig.11(b)) above atmospheric

$$= \frac{2\gamma}{R_1} = \frac{2\gamma}{r_1}\cos 40 = \frac{2 \times 0{\cdot}46 \times \cos 40}{10^{-4}}$$

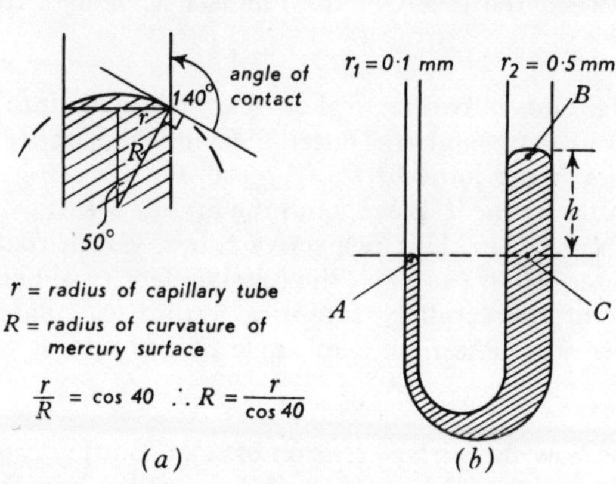

Fig. 11

Similarly the excess pressure at B above atmospheric

$$= \frac{2\gamma}{R_2} = \frac{2\gamma \cos 40}{r_2} = \frac{2 \times 0{\cdot}46 \times \cos 40}{5 \times 10^{-4}}$$

Hence the difference between the pressures at A and B

$$= 2 \times 0{\cdot}46 \times 10^4 \times \cos 40\,(1 - \tfrac{1}{5}) = 5{\cdot}6 \times 10^3 \text{ N m}^{-2} \qquad \text{(i)}$$

Now the pressure at A = pressure at C (on same horizontal level in the mercury), and the pressure difference between A and C (and, therefore, between A and B)

$$= g\,\rho h \qquad (\rho = \text{density of mercury})$$

$$= 9{\cdot}8 \times 1{\cdot}36 \times 10^4 \times h \qquad \text{(ii)}$$

Hence, equating (i) and (ii) we have

$$9{\cdot}8 \times 1{\cdot}36 \times 10^4 \times h = 5{\cdot}64 \times 10^3$$

giving $$h = \frac{5{\cdot}64 \times 10^3}{9{\cdot}8 \times 1{\cdot}36 \times 10^4}$$
$$= \underline{0{\cdot}0423} \text{ m}$$

19 A U-tube is made up of two lengths of capillary tubing of internal radii 1·0 mm and 0·2 mm respectively. The tube is partially filled with a liquid of surface tension 0·022 N m^{-1} and zero angle of contact, and when the tube is held in a vertical position a difference of 2·1 cm is observed between the levels of the menisci. Calculate the density of the liquid.

20 The lower ends of two vertical capillary tubes dip into two beakers, one containing water and the other benzene, the upper ends of the capillary tubes being joined by a T-piece. On reducing the pressure in the apparatus via the T-piece, both the liquids are drawn up through a height of 14·6 cm in their respective tubes. Given that the surface tension of water is 0·072 N m^{-1}, the relative density of benzene is 0·88, the diameter of the capillary tubing is 1 mm, calculate the surface tension of benzene. (Assume zero angle of contact for both benzene and water.)

21 Describe how the surface tension of soap solution can be obtained by a method involving the excess pressure inside a soap bubble.

A T-piece is provided with taps so that soap bubbles can be separately blown at either end of the cross-tube. Of two such bubbles blown, A has twice the radius of the other bubble B. Describe and explain what happens when A and B are put into communication by opening the taps between them.

22 What is the excess pressure inside a soap bubble of radius 5 cm if the surface tension of soap solution is 0·04 N m^{-1}? Find also the work done in blowing the bubble.

Worked example

23 *Two spherical soap bubbles, A and B, of radii* 3 *cm and* 5 *cm, coalesce so as to have a portion of their surfaces in common. Calculate the radius of curvature of this common surface.*

Let the radius of curvature of the common surface of the two soap bubbles be r_3 (see diagram). Then, since the excess pressure in bubble A above atmospheric pressure $= \frac{4\gamma}{r_1} = \frac{4\gamma}{3}$, and the excess pressure in bubble B above atmospheric pressure $= \frac{4\gamma}{r_2} = \frac{4\gamma}{5}$, we see that there is an excess pressure on the A-side of the common surface above that on the B-side of $\frac{4\gamma}{3} - \frac{4\gamma}{5}$. But this excess pressure must equal $\frac{4\gamma}{r_3}$.

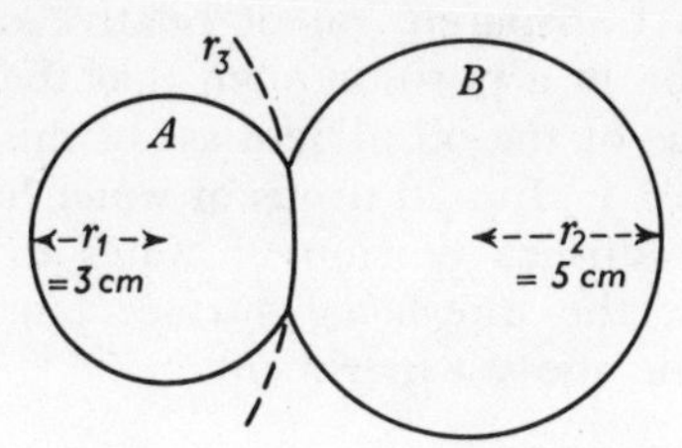

Fig. 12

Hence we have $$\frac{4\gamma}{r_3} = \frac{4\gamma}{3} - \frac{4\gamma}{5}$$

or $$\frac{1}{r_3} = \frac{1}{3} - \frac{1}{5}$$

giving $$r_3 = \underline{7{\cdot}5 \text{ cm}}$$

24 Obtain an expression for the excess pressure inside a spherical air bubble of radius r blown inside a liquid of surface tension γ.

Calculate the pressure inside a spherical air bubble of diameter 0·1 cm blown at a depth of 20 cm below the surface of a liquid of density $1{\cdot}26 \times 10^3$ kg m^{-3} and surface tension 0·064 N m^{-1}. (Height of mercury barometer = 0·76 m, density of mercury = $13{\cdot}6 \times 10^3$ kg m^{-3}.)

25 Obtain a general expression for the difference of pressure between the two sides of a liquid surface in terms of the surface tension of the liquid and the principal curvatures of the surfaces.

A soap bubble is formed over two wire rings, of radius 2 cm, placed parallel to each other. The distance between the rings is adjusted until the film between them is cylindrical. What is then the radius of curvature of the convex spherical caps on the ends?

26 A drop of water is placed between two glass plates which are then pushed together until the water drop is squashed out into a thin circular film of radius 5 cm and thickness 0·2 mm. Calculate the force, applied normal to the plates, which is required to separate them if the surface tension of water is 0·072 N m^{-1}. Give the underlying theory of your calculation.

27 Give the theory underlying the determination of the surface tension of a liquid by the 'drop weight' method.

Water, contained in a tank, is allowed to escape slowly via a length of clean vertical tubing. A beaker placed below this tubing is found to increase in mass by 6·9 g after 20 drops have fallen into it. A second beaker, containing a transparent oil of relative density 0·92 is now placed below the tube in a position such that the end of the tube is just below the surface of the oil. The mass of this beaker is found to have increased by 25·1 g after 20 drops of water have issued from the tube. Assuming the surface tension of water to be 0·072 N m^{-1}, calculate a value for the interfacial surface tension between water and the oil used in the above experiment.

28 Define the property of *surface tension* of a liquid. How is this property accounted for on the molecular theory of matter?

A capillary tube of length 5 cm and internal diameter 0·01 cm is mounted about an axis through one end perpendicular to the length of the tube and so as to allow the tube to be spun in a horizontal plane. If originally the tube was full of water, find the angular speed of the tube when exactly half of the water remains in the tube. (Surface tension of water $= 0{\cdot}7 \times 10^{-1}$ N m^{-1}, density of water $= 10^3$ kg m^{-3}.)

Viscosity

29 Distinguish between *orderly* and *turbulent* flow of a liquid and describe some experiment whereby you could demonstrate the difference between the two types of flow in the case of a liquid.

Water flows through a horizontal tube of length 25 cm and of internal diameter 1 mm under a constant head of the liquid 15 cm high. If a volume of 72·5 cm^3 of water issues from the tube in 10 minutes, calculate the coefficient of viscosity of water under the conditions of the experiment. (You may assume that conditions for orderly flow exist.)

30 Describe, with full experimental detail, a method for finding the coefficient of viscosity of a liquid such as water. Mention any special precautions which must be taken to ensure an accurate result.

A tank empties through a length of horizontal capillary tubing inserted near its base. After being filled with water the tank empties itself in 100 seconds, whereas on substituting turpentine for the water the time taken for the tank to empty is 165 seconds. If the relative density of turpentine is 0·87, and the coefficient of viscosity of water is 0·0012 N m^{-2} s, find that of turpentine.

31 Define *viscosity, coefficient of viscosity.*

Two thin-walled co-axial cylinders have radii of 5 cm and 5·2 cm, the inner one being capable of free rotation inside the larger cylinder. Water is now poured into the space between the cylinders to a depth of 10 cm. Find the approximate value of the retarding force on the curved surface of the inner cylinder on being rotated at a steady rate of 5 revolutions per minute. Viscosity of water at the temperature prevailing = 0·0012 N m^{-2} s.

32 Describe some form of viscosimeter suitable for the determination of the viscosity of a liquid such as glycerine and comment on any precautions or corrections to obtain a reliable result.

The times (t) of the steady axial descent of steel spheres, 3 mm in diameter, between two marks 15 cm apart scratched on the surface of vertical glass cylindrical tubes of diameter D containing a viscous liquid are set out in the table below. Use this data to find the velocity

t	7·5	7·2	6·6	6·3	6·1	5·9	s
D	2·0	2·5	3·5	5·0	8·0	10·0	cm

of descent for an infinite extent of the liquid and calculate its coefficient of viscosity at the temperature of the experiment given that the relative density of the liquid is 1·26 and that of steel is 7·80.

33 Define *critical velocity, terminal velocity.*

A small air bubble, released at the bottom of a jar containing olive oil, rises to the surface with a constant velocity of 1·2 cm s^{-1}. Given that the coefficient of viscosity of olive oil is 0·084 N m^{-2} s and that its relative density is 0·92, calculate the radius of the air bubble.

Worked example

34 *Small metal spheres of diameter* 2 *mm and density* $7{\cdot}8\times10^3$ *kg* m^{-3} *are found to fall through glycerine with a terminal velocity of* 0·006 *m* s^{-1}. *Calculate the coefficient of viscosity of glycerine given that its density is* $1{\cdot}26\times10^3$ *kg* m^{-3}.

The forces acting on a metal sphere when falling in the glycerine are:
Downwards: force of gravity

$$= \tfrac{4}{3}\pi(0{\cdot}001)^3 \times 7{\cdot}8\times10^3\times9{\cdot}8 \text{ N}$$

Upwards: Upthrust (U) of glycerine + viscous drag (F) of glycerine

$$= \tfrac{4}{3}\pi(0{\cdot}001)^3 \times 1{\cdot}26\times10^3\times9{\cdot}8 + F \text{ N}$$

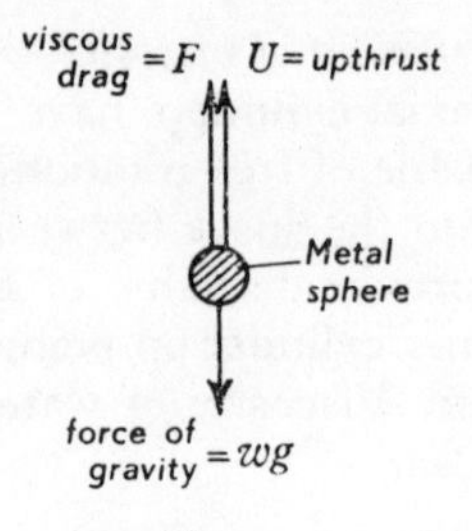

Fig. 13

When falling at a constant velocity (terminal velocity), there is no net force on the sphere, hence

$$F+\tfrac{4}{3}\pi(0{\cdot}001)^3\times 1{\cdot}26\times 10^3\times 9{\cdot}8 = \tfrac{4}{3}\pi(0{\cdot}001)^3\times 7{\cdot}8\times 10^3\times 9{\cdot}8$$

$$\text{or} \qquad F = \tfrac{4}{3}\pi(0{\cdot}001)^3(7{\cdot}8-1{\cdot}26)10^3\times 9{\cdot}8$$

Now by Stokes' law, $F = 6\pi\eta av$
where η is the coefficient of viscosity of the medium, a the radius of the sphere and v its terminal velocity,

$$\text{i.e.} \qquad F = 6\pi\eta(0{\cdot}001)\times 0{\cdot}006 \text{ N}$$

Hence we have

$$6\pi\eta(0{\cdot}001)\times 0{\cdot}006 = \tfrac{4}{3}\pi(0{\cdot}001)^3(7{\cdot}8-1{\cdot}26)10^3\times 9{\cdot}8$$

from which

$$\eta = \frac{2}{9}\frac{(0{\cdot}001)^2\times 6{\cdot}54\times 10^3\times 9{\cdot}8}{0{\cdot}006}$$

$$= \underline{2{\cdot}38} \text{ N m}^{-2}\text{ s}$$

35 Given that the viscous drag F for a particle of radius a and density ρ moving with a velocity v in a medium of viscosity η is given by $6\pi\eta av$, obtain an expression for the terminal velocity for such a particle when falling vertically through the medium.

A quantity of powdered chalk, containing particles of different sizes, is stirred up in a beaker of water. Assuming the particles to be spherical in shape, find the radius of the particles remaining in suspension 12 hours later if the depth of the water in the beaker is 10 cm. Take the density of chalk as $2{\cdot}6\times 10^3$ kg m^{-3} and the viscosity of water as $0{\cdot}0012$ N m^{-2}s.

36 When a potential of 7500 volt is applied across two horizontal plates situated 2 cm apart, an oil drop, carrying two free electrons, is found to fall between the plates at the steady rate of 0·2 mm s^{-1}. On reversing the electric field the drop rises at the steady rate of 0·1 mm s^{-1}. Calculate the radius of the drop if the electronic charge is $1{\cdot}59 \times 10^{-19}$ C and the viscosity of air is $1{\cdot}8 \times 10^{-5}$ N m^{-2} s.

37 Give a description of the method of determining the electronic charge by experiments on oil drops.

A charged oil drop, falling under gravity between two horizontal metal plates 3 cm apart, was observed to descend at a steady rate of 0·12 mm s^{-1}. When a potential difference of 8000 volt was applied across the plates, the descent of the oil drop was arrested. Calculate (*a*) the radius of the drop, (*b*) the charge on it.

(Densities of oil and air on $0{\cdot}93 \times 10^3$ kg m^{-3} and 1·3 kg m^{-3} respectively; viscosity of air $1{\cdot}83 \times 10^{-5}$ N m^{-2} s.)

38 Inserted into the lower end of a deep cylindrical glass vessel of diameter 10 cm is a horizontal capillary tube 30 cm in length with an internal diameter of 1 mm. The vessel is filled with water which is allowed to flow out via the capillary tube. Calculate the time for the level of the water to fall from a height of 30 cm to a height of 10 cm above the axis of the capillary tube given that the coefficient of viscosity of water = 0·0012 N m^{-2}s and $\log_e 10 = 2{\cdot}303$.

39 A thin uniform metal disc of mass 50 g and radius 5 cm is pivoted so that it can freely rotate between two metal plates mounted on either side of the disc with a clearance of 2 mm.

If this arrangement is submerged in a tank of water, find how long it will take for the angular velocity of the disc to fall to one-fifth of its initial value on being set rotating. Viscosity of water = 0·0012 N m^{-2} s.

Gravitation

40 State *Newton's law of universal gravitation.*

Derive a relationship between the constant of gravitation and the mean density of the earth. Describe some laboratory method whereby the values of these quantities have been found.

41 Two small lead spheres, each of mass 20 g, are suspended side by side by threads 20 metres long, the upper ends of the threads being 3 cm apart. Find, approximately, how much less than 3 cm is the distance between the centre of the lead spheres. Take G as $6{\cdot}7 \times 10^{-11}$ N m^2 kg^{-2} and g as 9·80 m s^{-2}.

Worked example

42 *Compare the value of the acceleration due to gravity on the surface of Mercury with its value on the Earth's surface given that: Radius of Mercury = 0·38 × radius of Earth, mean density of Mercury = 0·68 × mean density of Earth.*

The force of attraction of the Earth (mass M, radius R) on a body of mass m near its surface is, by Newton's law of gravitation,

$$G\frac{Mm}{R^2} \quad (G \text{ being the constant of gravitation}).$$

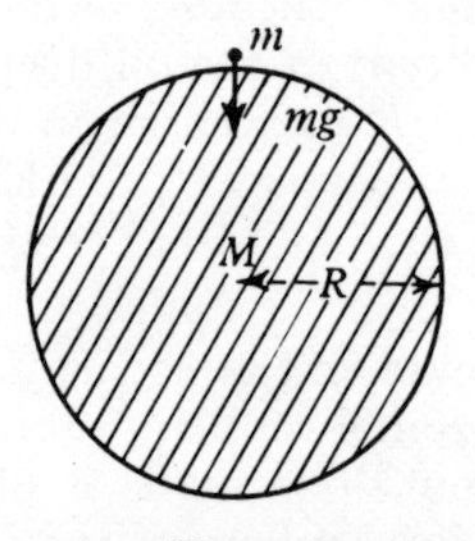

Fig. 14

Now if g is the acceleration of gravity at the surface of the planet, the force on a mass m kg is mg N, and therefore

$$mg = G\frac{Mm}{R^2}$$

or $g = G\dfrac{\frac{4}{3}\pi R^3\Delta}{R^2} = \frac{4}{3}\pi R\Delta$ (Δ being the mean density of the planet).

Hence, using the subscripts M and E to refer to Mercury and Earth respectively, we have

$$\frac{g_M}{g_E} = \frac{R_M\Delta_M}{R_E\Delta_E} = \left(\frac{R_M}{R_E}\right)\cdot\left(\frac{\Delta_M}{\Delta_E}\right)$$
$$= 0{\cdot}68 \times 0{\cdot}38 = \underline{0{\cdot}26}$$

43 Obtain an expression for the acceleration of gravity at a height h above the Earth's surface in terms of the radius R of the Earth and surface value g of the acceleration of gravity. (Consider only the case where h is small compared with R.)

If a pendulum has a periodic time of exactly 1 second at the Earth's surface, what would be its period 10 km above the Earth's surface? Take the radius of the Earth as 6400 km.

44 Define the *constant of gravitation* and describe a non-laboratory experiment by means of which its value has been determined.

Assuming the orbits of the planets to be circular, calculate the radius of the orbit of Mars if that of the Earth is $1{\cdot}496 \times 10^8$ km and the periods of Mars and the Earth are 687 days and 365 days respectively.

45 State *Kepler's laws of planetary motion* and show how Newton made use of them in deducing his law of gravitation.

Calculate the mean distance of the Moon from the Earth if its period of rotation round the Earth is 27·3 days and the radius of the Earth is 6376 km. Take the acceleration due to gravity at the Earth's surface $9{\cdot}80$ m s^{-2}.

46 Give an account of the variation of the acceleration due to gravity over the Earth's surface.

If the polar value of the acceleration due to gravity is $9{\cdot}832$ m s^{-2}, calculate its value for a place in latitude 45° assuming the Earth to be a true sphere of radius 6370 km.

47 Give an account of Airey's mine experiment to determine the mean density of the Earth.

Compare the value of the acceleration of gravity at the surface of the Earth with that at a point 5 km deep in the Earth's crust given that the density of the crust $= 2{\cdot}5 \times 10^3$ kg m^{-3}, the mean density of the Earth $= 5{\cdot}5 \times 10^3$ kg m^{-3}, and the radius of the Earth = 6400 km.

48 Show that the rotation of the Earth causes a plumb-line to hang slightly out of the vertical in all latitudes except 0° and 90°. Show further that this effect is a maximum for latitude 45° and calculate its magnitude for this latitude if g is $9{\cdot}81$ m s^{-2} and the radius of the Earth is 6400 km.

49 If a vertical tunnel were to be bored right through the Earth so as to pass through its centre, show that the subsequent motion of a stone dropped into the tunnel would be simple harmonic.

Find the velocity with which the stone would pass through the centre position and calculate the periodic time of the motion. ($G = 6{\cdot}66 \times 10^{-11}$ N m^2 kg^{-2}; mean density of Earth $= 5{\cdot}5 \times 10^3$ kg m^{-3}; radius of Earth = 6400 km.)

Rocketry and satellite motion

50 Find the minimum horizontal velocity with which a body must be projected from a place on the earth's surface in order that the body may revolve as a satellite just clear of the earth's surface. Take the radius of the Earth as $6{\cdot}4 \times 10^3$ km.

51 State the forces acting on an earth satellite while in orbit and explain why it maintains its orbit (assumed circular and concentric with the Earth's centre).

Given that the radius of the earth is 6400 km and the value of the acceleration of gravity at its surface is $9{\cdot}8$ m s^{-2}, calculate the orbital time of a small satellite which *just* clears the surface of the Earth.

52 Obtain an expression in terms of the radius of the Earth R and the acceleration of gravity g at the Earth's surface for the orbital time of a satellite orbiting the Earth at a constant height h above the Earth's surface.

Using the data of the previous problem calculate the height above the Earth's surface at which a satellite will have exactly twice the orbital time as one that just clears the earth's surface.

53 A satellite revolves in a circular orbit round the Earth in the plane of the equatorial section. What is its height above the Earth's surface if, to an observer on the equator, the satellite appears to be constantly overhead? What, also, is the satellite's speed at this height?

(Radius of Earth = $6{\cdot}4 \times 10^6$ m, acceleration of gravity at Earth's surface = $9{\cdot}8$ m s^{-2}.)

54 Obtain an expression for the minimum velocity which a body must have if it is to escape fully from the Earth's gravitational field. Give your expression in terms of the gravitational constant G and the mass M and radius R of the Earth.

Find this value given that:
$G = 6{\cdot}67 \times 10^{-11}$ Nm^2 kg^{-2}, $M = 5{\cdot}98 \times 10^{24}$ kg, $R = 6400$ km.

55 Using the data of the above problem, and given that the mass of the moon is $\frac{1}{81} \times$ that of the Earth, and the Moon's radius is $0{\cdot}27 \times$ that of the Earth, calculate the corresponding value of the above velocity from the moon's surface.

In the light of these calculated velocities, and given that the velocities at 0°C of hydrogen, helium, nitrogen and oxygen molecules are respectively $1{\cdot}84$, $1{\cdot}31$, $0{\cdot}495$ and $0{\cdot}461 \times 10^5$ m s^{-1}, comment on the presence of these gases (*a*) in the Earth's atmosphere, (*b*) the Moon's.

56 A manoeuvre or change in velocity of a space vehicle involves discharging mass in the form of rocket exhaust. By applying the conservation principle to the total momentum of the system show that, for a total initial mass M_0 and a constant exhaust velocity relative to the vehicle of v_e, the increase of velocity Δv of the space vehicle when it fires off a mass of ΔM of exhaust fuel is given by the expression

$$\Delta v = v_e \log_e \left(\frac{M_0}{M_0 - \Delta M} \right)$$

Derive also expressions for (i) the thrust given to the rocket and (ii) the minimum power required for the manoeuvre.

57 Define *gravitational potential* and determine the energy needed to lift unit mass from the Earth's surface completely clear of its gravitational pull.

A space rocket is to be projected fully clear of the Earth into outer space. If 75 per cent of the mass of the rocket consists of fuel which has a constant exhaust velocity of v_e relative to the rocket, find the value of v_e for the rocket to be projected in the manner described assuming that the fuel is fully burnt in the early stages of flight close to the Earth. Take the radius of Earth as 6400 km and the acceleration of gravity at the Earth's surface as $9{\cdot}8 \text{ m s}^{-2}$.

Method of dimensions

58 Explain clearly what you understand by (*a*) *fundamental units,* (*b*) *derived units,* (*c*) the *dimensions* of a physical quantity.

Express in dimensional form the following physical quantities: (i) acceleration, (ii) force, (iii) surface tension, (iv) coefficient of viscosity, (v) the constant of gravitation.

59 Specify the physical quantities on which the frequency of a tuning fork may reasonably be supposed to depend and apply the method of dimensions to obtain an equation connecting the various quantities involved.

Worked example

60 *After being deformed a spherical drop of liquid will execute periodic vibrations about its spherical shape. Using the method of dimensions, obtain an expression for the frequency of these vibrations in terms of the related physical quantities.*

The frequency (f) of vibrations of the drop will depend on the surface tension (γ) of the drop, its density (ρ) and on the radius (r) of the drop. Hence we may write

$$f = k\gamma^x \rho^y r^z \qquad (1)$$

where k is some constant.

To obtain the values of x, y and z we must insert the dimensions of the various quantities involved in equation (1). This gives us the dimensional equation,

$$[T^{-1}] = [MT^{-2}]^x[ML^{-3}]^y[L]^z$$

k, being a numeric, has no dimensions.

Equating the indices of the dimensions we have,

For the dimension of M: $0 = x+y$
,, ,, ,, ,, L: $0 = 3y+z$
,, ,, ,, ,, T: $-1 = -2x$

From these equations we get

$$x = \tfrac{1}{2}, \quad y = -\tfrac{1}{2}, \quad z = -\tfrac{3}{2}$$

Hence
$$f = k\gamma^{\frac{1}{2}}\rho^{-\frac{1}{2}}r^{-\frac{3}{2}}$$
$$= k\sqrt{\frac{\gamma}{\rho r^3}}$$

which is the required formula.

61 A thin circular metal disc of radius 4 cm is pivoted about a central axis perpendicular to its plane and is arranged to spin when completely submerged in a liquid of viscosity 0·0012 N m^{-2} s. If it takes 10 seconds for the angular velocity of the disc to fall to half its value, calculate the time for a similar shrinkage of the angular velocity of a disc of radius 10 cm made from the same metal sheet but rotating in a liquid of viscosity 0·0004 N m^{-2} s.

62 Explain the *principle of dimensional homogeneity*. Discuss its power and its limitations.

Apply the principle to show how the velocity of the transverse vibrations of a stretched string depend on its length (l), mass (m) and the tensional force (F) in the string.

63 As the pressure gradient along a capillary tube increases, the velocity of orderly flow of a liquid through the tube increases until, at a certain critical velocity, turbulence sets in. Use the method of dimensions to obtain a relation between this critical velocity and the viscosity of the liquid, its density and the radius of the tube—the quantities on which it depends.

If the critical velocity for water (viscosity 0·0012 N m^{-2} s) using a tube of 1 mm diameter is 2·4 m s^{-1}, find that for mercury (viscosity 0·0016 N m^{-2} s, density $13{\cdot}6 \times 10^3$ kg m^{-3}) for a similar tube.

64 Show, by the method of dimensions, that when a body is moving through a fluid under conditions such that the resisting force is proportional to the square of the velocity of the body, then the resisting force is independent of the viscosity of the liquid.

HEAT

Some data and useful constants

Specific heat capacity of water	$= 4{\cdot}185 \times 10^{3}$ J kg^{-1} K^{-1}
Specific heat capacity of copper	$= 0{\cdot}381 \times 10^{3}$ J kg^{-1} K^{-1}
Specific latent heat of ice	$= 0{\cdot}334 \times 10^{6}$ J kg^{-1}
Specific latent heat of steam	$= 2{\cdot}243 \times 10^{6}$ J kg^{-1}
Thermal conductivity of copper	$= 3{\cdot}84 \times 10^{2}$ W m^{-1} K^{-1}
Molar volume at s.t.p. (Vm)	$= 2{\cdot}24136 \times 10^{-2}$ m^{3} mol^{-1}
Molar gas constant (R)	$= 8{\cdot}3143$ J mol^{-1} K^{-1}
Avogadro constant (L)	$= 6{\cdot}02252 \times 10^{23}$ mol^{-1}
Standard atmosphere ($\equiv$ 760 mm of mercury)	$= 101325$ N m^{-2}
Stefan constant (σ)	$= 5{\cdot}6697 \times 10^{-8}$ W m^{-2} K^{-4}
Absolute zero of temperature	$= -273{\cdot}15$ (exactly) °C

Thermometry

1 What is meant by a *scale of temperature*? Discuss the properties of a suitable thermometric substance, indicating how these properties are used to define a scale of temperature.

2 Discuss the use of mercury as the thermometric substance in liquid-in-glass thermometers. In what way have the range of such thermometers been extended?

3 Summarize the defects of the mercury-in-glass thermometer which limit its use in accurate scientific work.

4 Describe some form of constant volume gas thermometer and explain how you would use it to determine the boiling point of a given liquid.

Such a thermometer is used to determine the temperature of a furnace when it is found that the excess pressure in the bulb over atmospheric pressure is 205 cm of mercury. With the bulb of the thermometer in melting ice the air pressure in the bulb is 10 cm of mercury below atmospheric which is constant at 76 cm of mercury throughout the experiment. Calculate the temperature of the furnace. What assumptions have you made in your calculation?

5 What is meant by the *fundamental interval* of a thermometric scale? How is this interval used to define temperature on the scale?

The numerical value of the physical property of a given substance is 1·05 at the ice point and 1·77 at the steam point. At what temperature will the numerical value of the physical property be 1·21?

6 Discuss the general methods used in the measurement of temperature and indicate the importance of expressing the results in terms of a standard scale.

The volume of a certain liquid at different temperatures is given by the expression

$$V_t = V_0 (1 + \alpha t + \beta t^2)$$

where $\alpha = 0{\cdot}0011$, $\beta = -0{\cdot}000\,002$ and t is the temperature measured on the constant volume gas scale. If a thermometer, graduated on the Centigrade (Celsius) scale, is constructed using this liquid, what temperature will it record when $t = 40$°C?

7 Describe the details of construction of the platinum resistance thermometer.

The resistance of a given wire at various temperatures on the constant volume gas scale are as under:

t°C	0	10	20	30	40	50	60	70	80	90	100
Resistance (ohm)	5·00	5·08	5·16	5·23	5·31	5·40	5·50	5·61	5·73	5·86	6·00

Find (*a*) the temperature on the resistance scale corresponding to 75°C on the gas scale, (*b*) the temperature on the gas scale corresponding to 35°C on the resistance scale.

Expansion of solids

8 Discuss in detail **two** applications of the expansion of metals.

Calculate the lengths of brass rod and iron rod such that they may have a constant difference of length of 5 cm at all temperatures. Coefficients of linear expansivity of brass and iron are 0·000018 K^{-1} and 0·000012 K^{-1} respectively.

9 An iron ring of internal diameter 29·95 cm is to be fitted on a wooden cylinder of diameter 30·00 cm. Find the range of temperature through which the ring must be heated in order that this is just possible. Coefficient of linear expansivity of iron = 0·000012 K^{-1}.

Worked example

10 *A length of copper wire 1 mm in diameter at room temperature is to be passed through a circular hole in an iron plate. What must be the diameter of this hole, at room temperature, for the area of the annular aperture surrounding the wire to be constant at all temperatures? Coefficient of linear expansivity of copper* $= 17 \times 10^{-6}$ K^{-1}, *of iron* $= 12 \times 10^{-6}$ K^{-1}.

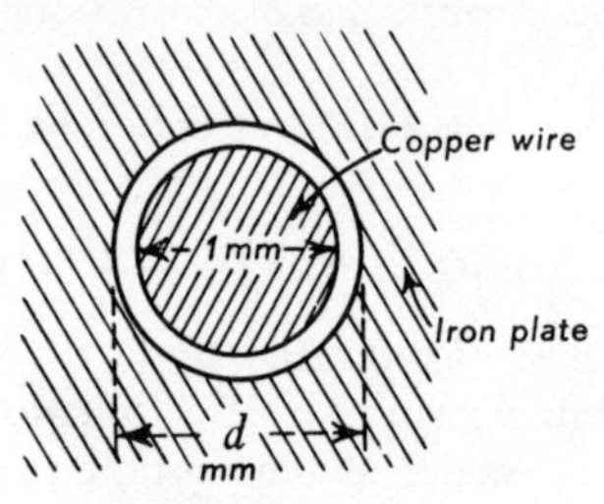

Fig. 15

Let the diameter of the circular aperture at room temperature $= d$ mm. Then area of annular aperture at room temperature

$$= \frac{\pi d^2}{4} - \frac{\pi}{4} = \frac{\pi}{4}(d^2 - 1) \text{ mm}^2$$

If the temperature rises by t°C, the new diameters of wire and aperture are respectively

$$(1 + 17 \times 10^{-6} t) \quad \text{and} \quad d(1 + 12 \times 10^{-6} t) \text{ mm}$$

and accordingly the area of the aperture at this temperature is

$$\frac{\pi}{4} d^2(1 + 12 \times 10^{-6} t)^2 - \frac{\pi}{4}(1 + 17 \times 10^{-6} t)^2 \text{ mm}^2$$

$$= \frac{\pi}{4} d^2(1 + 24 \times 10^{-6} t) - \frac{\pi}{4}(1 + 34 \times 10^{-6} t) \quad \text{ignoring the terms in } t^2\text{, etc.}$$

$$= \frac{\pi}{4}(d^2 - 1) + \frac{\pi}{4}(d^2 \times 24 \times 10^{-6} t - 34 \times 10^{-6} t)$$

Hence, for the aperture area to be constant,

$$\frac{\pi}{4}(d^2 \times 24 \times 10^{-6} t - 34 \times 10^{-6} t) = 0$$

or $\quad 12d^2 - 17 = 0$

from which

$$d = \sqrt{\frac{17}{12}} = \underline{1{\cdot}19} \text{ mm}$$

11 Calculate the percentage change in the moment of inertia of a steel flywheel (which you may assume to be a uniform circular disc) due to a rise of temperature of 10°C. (Coefficient of linear expansivity of steel $= 1{\cdot}1 \times 10^{-5}\ \mathrm{K}^{-1}$.)

12 A clock with a brass pendulum keeps correct time when the temperature is 10°C. What will be the error in the time recorded by the clock if it is placed for one week in a room at a constant temperature of 20°C? (Coefficient of linear expansivity of brass $= 2 \times 10^{-5}\ \mathrm{K}^{-1}$.)

13 What do you understand by the statement that the coefficient of linear expansivity of steel is $1{\cdot}1 \times 10^{-5}\ \mathrm{K}^{-1}$?

The ends of a cylindrical steel rod of diameter 1 cm are rigidly held between two firm clamps. Find the force the clamps must exert on the rod to prevent it from expanding when its temperature is raised by 20°C. (Young's modulus for steel $= 2 \times 10^{11}\ \mathrm{N\ m^{-2}}$, acceleration of gravity $= 9{\cdot}80\ \mathrm{m\ s^{-2}}$.)

14 A trigger switch depends for its action on the differential expansion of a compound strip of brass and iron each 1 mm thick. The two strips are rigidly fixed together along their length, and when cold are each 10 cm long. One end of the compound strip is firmly secured, whilst the free end is 1 mm above an electrical contact point. If the thermal capacity of the bimetallic strip is $8\ \mathrm{J\ K^{-1}}$, find the time elapsing before the electric circuit is made if heat at the rate of $20\ \mathrm{J\ s^{-1}}$ is supplied to the strip. (Coefficients of linear expansivity of brass $= 0{\cdot}000018\ \mathrm{K}^{-1}$, of iron $= 0{\cdot}000\,012\ \mathrm{K}^{-1}$.)

Expansion of liquids

15 A loaded cylindrical glass tube, provided with a centimetre scale reading from the bottom of the tube, floats vertically in water at the 20·0 cm mark when the temperature of the water is 10°C. What will be the temperature of the water when the tube floats at the 20·1 cm mark? (Coefficient of cubic expansivity of water $= 0{\cdot}00042\ \mathrm{K}^{-1}$, coefficient of linear expansivity of glass $= 0{\cdot}000\,008\ \mathrm{K}^{-1}$.)

16 A cylindrical glass tube contains mercury at 0°C. What will be the percentage change (*a*) in the height of the mercury column, (*b*) in the pressure exerted by the mercury at the bottom of the tube, on raising the temperature by 50°C? (Coefficient of cubic expansivity of glass and mercury are $0{\cdot}000\,024$ and $0{\cdot}000\,18\ \mathrm{K}^{-1}$ respectively.)

17 A solid or relative density 0·88 floats in a beaker of olive oil of relative density 0·92. The coefficient of linear expansivity of the solid is $1{\cdot}2 \times 10^{-4}$ K^{-1} and the cubic expansivity of olive oil is 7×10^{-4} K^{-1}. Through what range of temperature must the beaker and its contents be raised before the solid just sinks in the oil?

18 Define the *coefficients of real and apparent cubic expansivity* of a liquid and establish the relation between them.

A mercury thermometer is to be made with glass tubing of internal diameter 0·4 mm so that the distance between the upper and lower fixed points is 10 cm. Calculate what the internal volume of the bulb and stem below the lower fixed point must be in the finished thermometer.

Coefficient of cubic expansivity of mercury = 0·000 18 K^{-1}
Coefficient of linear expansivity of glass = 0·000 008 K^{-1}

Worked example

19 *The volume of the mercury in a mercury-in-glass thermometer is 0·50 cm^3 at 0°C and the distance between the upper and lower fixed points is 15·0 cm. What is the diameter of the bore of the stem of the thermometer?*

Coefficient of linear expansivity of glass = 0·000 008 K^{-1}.
Coefficient of cubic expansivity of mercury = 0·000 181 K^{-1}.

The coefficient of apparent cubic expansivity (γ_a) of the mercury

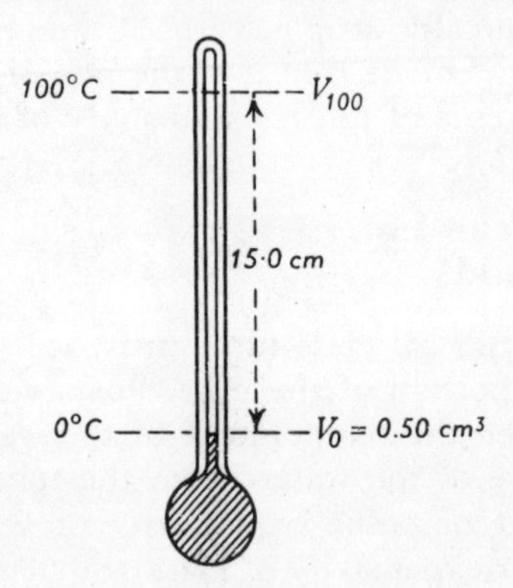

Fig. 16

against the glass = coefficient of cubic expansivity of the mercury − coefficient of cubic expansivity of glass

$$= 0{\cdot}000\,181 - 3 \times 0{\cdot}000\,008$$
$$= 0{\cdot}000\,157 \text{ K}^{-1}$$

Now if V_0 = the internal volume of the thermometer up to the 0°C marking, and V_{100} the internal volume up to the 100°C marking, then, by the expansion law $V_t = V_0(1+\gamma_a t)$ we have

$$V_{100} = V_0(1+\gamma_a t)$$

or $$V_{100}-V_0 = V_0\gamma_a t = 0{\cdot}5 \times 0{\cdot}000\,157 \times 100 \text{ cm}^3$$

But $$V_{100}-V_0 = \frac{\pi d^2}{4} \times 15$$

when d is the diameter of the bore in cm.

Hence $$\pi d^2 \times \frac{15}{4} = 0{\cdot}5 \times 0{\cdot}000\,157 \times 100$$

from which $$d = \sqrt{\frac{4 \times 0{\cdot}5 \times 0{\cdot}000\,157 \times 100}{15\pi}}$$

$$= \underline{0{\cdot}026 \text{ cm}}$$

20 What would be the volume at 0°C of the alcohol in an alcohol-in-glass thermometer with a stem of the same bore as the thermometer in the question above if the two thermometers are to be equally sensitive? (Coefficient of cubic expansivity of alcohol = 0·001 2 K^{-1}.)

21 Describe the volume dilatometer and critically discuss its use for the measurement of the coefficient of cubic expansivity of a liquid.

The glass bulb of a dilatometer has a volume of 5 cm^3. Calculate what volume of mercury must be put into the bulb in order that the residual space in it will not vary with temperature.

Coefficient of linear expansivity of glass = 0·000 008 K^{-1}
Coefficient of cubic expansivity of mercury = 0·000 18 K^{-1}

22 Give an account of the corrections for temperature to be made to the readings of a mercurial barometer.

Such a barometer is provided with a brass scale the markings on which are correct at 0°C. On a day when the air temperature is 20°C this barometer reads 74·865 cm. What is the true reading at this temperature, and what will the barometer reading be when reduced to 0°C? (Coefficient of linear expansivity of brass = 0·000 019 K^{-1}, coefficient of cubic expansivity of mercury = 0·000 181 K^{-1}.)

23 A glass bottle is filled with a given liquid at 0°C and, when its temperature is raised to 40°C, 0·12 g of the liquid is expelled. On now raising the temperature to 100°C a further 0·17 g of the liquid is expelled. Calculate (*a*) the mass of the liquid originally in the bottle, (*b*) the coefficient of real cubic expansivity of the liquid. Take the coefficient of cubic expansivity of glass as 0·000 01 K^{-1}.

24 Describe the weight thermometer method for the determination of the coefficient of cubic expansivity of a liquid.

When full of mercury a glass bottle has a mass of 600 g. If, on being heated through 40°C, 3·14 g of mercury are expelled, what is the mass of the bottle? (Coefficient of cubic expansivity of mercury = 0·000181 K^{-1}, coefficient of linear expansivity of glass = 0·000008 K^{-1}.)

25 Describe a method for determining the coefficient of apparent cubic expansion of a liquid.

A piece of glass weighs 3·350 N in air, 2·079 N when immersed in water at 4°C, and 2·127 N when the temperature of the water is raised to 100°C. Given that the coefficient of cubic expansivity of glass is 0·000024 K^{-1}, find the density of water at 100°C. Establish all formulae used and state any assumptions made in the calculation.

26 Describe and give the theory of a method for the direct determination of the coefficient of absolute cubic expansivity of a liquid.

One limb of a glass U-tube containing a liquid is surrounded with melting ice and the vertical height of the liquid in the other limb is 75·10 cm. Calculate the vertical height of the liquid in the other limb if it is maintained at a temperature of 80°C. (Coefficient of cubic expansivity of the contained liquid = 0·00094 K^{-1}.)

Expansion of gases. The gas laws

27 State Boyle's law and describe how its accuracy may be verified in the laboratory for air for pressures between $\frac{1}{2}$ and 2 atmospheres.

The pressure of air above the mercury in a barometer causes it to read 75 cm on a day when the true barometric height is 76 cm of mercury. Find how far the barometer tube must be depressed into the mercury cistern for the mercury in the tube to be 74 cm above the level outside it if originally the closed end of the tube was 80 cm above the cistern mercury level.

28 Give an account of the experimental procedures designed to test the validity of Boyle's law over extended ranges of pressure. Summarize the general nature of the results of such experiments, and briefly indicate the way in which the results have been interpreted.

29 A U-tube, containing mercury, has some air entrapped above the mercury in the closed limb. The other limb is open to the atmosphere, and when the level of the mercury in this limb is increased (by the addition of further mercury) from 10 cm below the level of the mercury in the closed limb to 10 cm above it, it is found that the length of the air column in the closed limb decreases from 15 cm to 11·5 cm. What is the pressure of the atmosphere (in cm of mercury)?

30 A cylindrical diving-bell of vertical height 4 m contains air at a temperature of 20°C and a pressure of 75 cm of mercury. The bell is then lowered into water at a temperature of 4°C until the water inside the bell rises half-way up the sides. Assuming no air escapes, calculate the depth of the top of the bell below the water surface. (Relative density of mercury = 13·6.)

Worked example

31 *A faulty barometer tube,* 80 *cm long, contains some air in the 'vacuum' space above the mercury and reads* 75 *cm when the true barometric height is* 76 *cm. What is the true atmospheric pressure on an occasion when the faulty barometer reads* 73 *cm, the temperature of the air being the same on both occasions?*

1*st occasion.* The pressure of the air will support a column of mercury 76 cm long. In the faulty barometer the mercury column is 75 cm long and hence the pressure of the air above the mercury is $76-75 = 1$ cm of mercury ($= p_1$). Assuming the tube to be of uniform cross-sectional area A cm^2, the volume of the air is $(80-75)A = 5A$ cm^3 ($= v_1$).

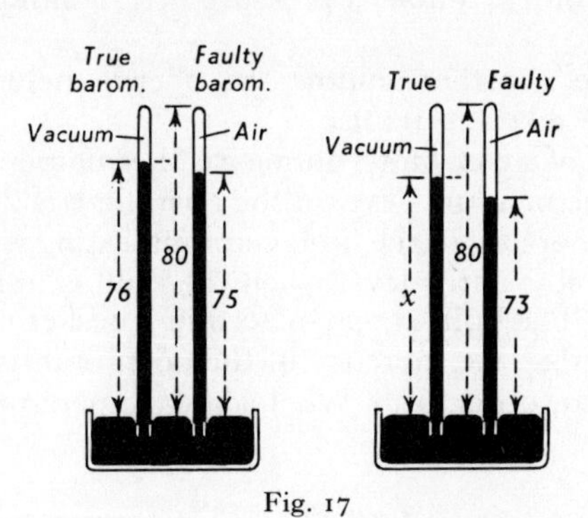

Fig. 17

2*nd occasion.* Let the mercury of a true barometer stand at x cm. The pressure exerted by the air in the faulty barometer is then $x-73$ ($= p_2$), and the volume of the air is $(80-73)A = 7A$ cm^3 ($= v_2$).

Then, since the temperature of the air is the same on both occasions, we have, applying Boyle's law to the mass of the air,

$$p_1v_1 = p_2v_2$$

i.e. $$1 \times 5A = (x-73) \times 7A$$

or $$5 = 7x - 511$$

giving $$x = \frac{516}{7} = \underline{73{\cdot}71}$$

for the true atmospheric pressure in cm of mercury.

32 Describe, with a suitable diagram, the construction and mode of action of a pump capable of producing low pressures.

The barrel of an exhaust pump has an effective volume of 100 cm^3 and is being used to extract air from a 1000 cm^3 flask. Ignoring the volume of the connecting tube, and assuming the temperature of the air to remain constant throughout, calculate the number of complete strokes of the pump needed to reduce the pressure of the air in the flask to one-hundredth of its initial value.

33 State 'Charles' Law' and describe a laboratory method of verifying the law.

When fully inflated with hydrogen gas at a temperature of 20°C and at a pressure of 1 atmosphere, a balloon has a volume of 10 000 m^3. What was the initial volume of this hydrogen gas if stored in cylinders at a temperature of 6°C under a pressure of 150 atmospheres?

34 Describe the constant-volume gas thermometer and state the precautions to be taken in its use.

When the bulb of a constant-volume gas thermometer is surrounded by melting ice, the mercury-level in the open limb is 2·3 cm below the fixed level of the mercury in the limb communicating with the gas bulb, but it is 20·4 cm above this level when the bulb is immersed in water boiling at 100°C. The bulb is now placed in a beaker of boiling liquid, when it is found that the mercury in the open limb is 10·2 cm above the fixed level in the other limb. What is the temperature of this boiling liquid?

35 Explain what is meant by (a) *the ideal gas equation,* (b) *the molar gas constant.*

Obtain a value for the molar gas constant using the data given below:

The molar volume at s.t.p. $= 2{\cdot}24 \times 10^{-2}$ m^3.

Standard pressure is the same as that exerted by a column of mercury 76 cm high.

Density of mercury $= 1{\cdot}36 \times 10^4$ kg m^{-3}.
Acceleration of gravity $= 9{\cdot}81$ m s^{-2}.

36 Define (a) *coefficient of increase of pressure at constant volume*, (b) *coefficient of increase in volume at constant pressure* and show that these coefficients are equal in the case of a perfect gas.

A glass bulb is fitted with a narrow tube open to the atmosphere. Calculate the fraction of the original mass of the air in the bulb which is expelled when the temperature of the bulb is raised from 0°C to 100°C.

Worked example

37 *Two glass bulbs, A and B, of volumes 500 cm³ and 200 cm³ respectively, are connected by a short length of capillary tubing and the apparatus, which is sealed, contains dry air at a pressure of 76 cm of mercury and at a temperature of 27°C. What does the pressure of the air in the apparatus become if the temperature of the larger bulb is raised to 127°C, the temperature of the other bulb remaining at 27°C?*

On heating A some air passes from A to B. Let v cm³ of the original

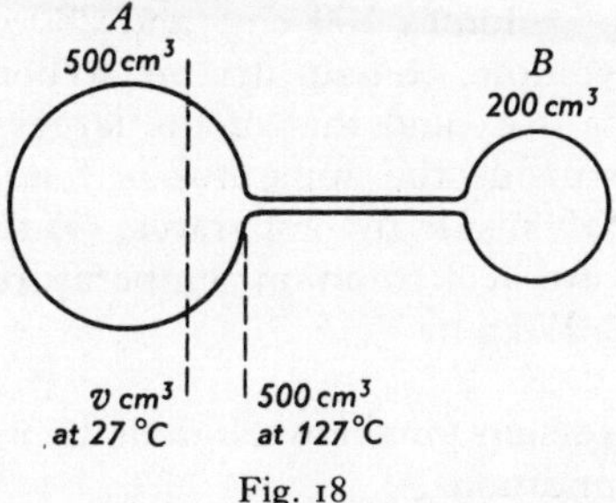

Fig. 18

volume (i.e. as measured at a temperature of 27°C) remain in A to fill it at the higher temperature. Then, if the air pressure is p cm of mercury, we have, applying the gas laws to this mass of gas in sphere A,

$$\frac{p \times 500}{400} = \frac{76 \times v}{300} \qquad (1)$$

Since $(500 - v)$ cm³ of air (at temperature 27°C) move across to B, the mass of the gas in B is increased in the proportion

$$\frac{(500 - v) + 200}{200} = \frac{700 - v}{200},$$

and hence, since the temperature remains constant, the pressure in B increases to

$$\left(\frac{700-v}{200}\right) \times 76 \text{ cm of mercury}$$

$$= p \text{ (the new pressure in the apparatus)} \qquad (2)$$

Re-arranging, we have

$$76v = 700 \times 76 - 200p \qquad (2a)$$

and eliminating v from equations (1) and (2a) we get

$$p \times \frac{500}{400} = \frac{700 \times 76 - 200p}{300}$$

from which $p = \dfrac{28 \times 76}{23} = \underline{92{\cdot}5}$ cm of mercury

38 Two glass bulbs, each of volume 100 cm^3, are in communication through a length of capillary tubing which has a bore of cross-sectional area 1 mm^2 and which contains a short mercury index. The bulbs contain air which is initially at a temperature of 27°C. What temperature change of either bulb will cause the index to move 1 mm along the capillary tube?

39 Two vessels, of volumes 100 cm^3 and 200 cm^3, connected by a tube of negligible volume, contain dry air. When the temperature of the smaller vessel is 0°C, and that of the larger vessel is 100°C, the pressure of the air inside the apparatus is 1 atmosphere. Calculate, (*a*) the mass of the air inside the apparatus, (*b*) the pressure of the air when both vessels are at a common temperature of 0°C. Density of dry air at s.t.p. = 1·29 kg m^{-3}.

40 Derive an expression for the work done by a gas when expanding under isothermal conditions.

Calculate the work done by 4 g of oxygen gas when expanding to three times its original volume at a constant temperature of 27°C. Density of oxygen at s.t.p. = 1·429 kg m^{-3}. Take the usual values for any other data required. ($\log_e$ N = 2·303 $\log_{10}$ N.)

Specific heat capacity

41 Define the term *specific heat capacity*.

When a block of metal of specific heat capacity 400 J kg^{-1} K^{-1} and mass 110 g is heated to 100°C and then quickly transferred to a calorimeter containing 200 g of a liquid at 10°C, the resulting temperature is 18·0°C. On repeating the experiment with 400 g of liquid in the same calorimeter, and at the same initial temperature of 10°C, the resulting temperature is 14·5°C. Calculate from these observations the specific heat capacity of the liquid and the heat capacity of the calorimeter.

42 Give a critical account of the method of mixtures for determining the specific heat capacity of a substance.

A mass of 100 g of a hot liquid is poured into 200 g of cold water contained in a beaker and a rise of temperature of 5°C is observed. The experiment is now repeated, but this time 200 g of water is heated and poured into the 100 g of liquid contained in the beaker when the temperature rise is observed to be 20°C. If the initial temperatures of the hot and cold liquid are the same in each experiment, obtain a value for the specific heat capacity of the liquid. (You may disregard the heat capacity of the beaker and assume no chemical reaction takes place between the liquids on mixing.) Take the specific heat capacity of water to be 4200 J kg^{-1} K^{-1}.

43 The initial temperatures of a mass m of liquid A, 2 m of liquid B, and 3 m of liquid C are respectively 30°C, 20°C and 10°C. On mixing liquids A and B the resulting temperature is 25°C; on mixing liquids B and C the resulting temperature is 14·5°C. What will be the resulting temperature on mixing liquids A and C? (Assume no chemical reaction between the liquids.)

44 When 100 g of a liquid at 100°C are mixed with 100 g of water at 15°C, the temperature of the mixture is 85°C. On repeating the experiment with the initial temperature of the liquid at 50°C (but with the same initial water temperature), the temperature of the mixture is 65°C. Explain these observations and make suitable deductions from them. (Assume no external heat losses and take the specific heat capacity of water as 4200 J kg^{-1} K^{-1}.

45 Two solid copper spheres of diameters 10 cm and 5 cm are at temperatures which are respectively 10°C and 5°C above that of the surroundings. Assuming Newton's law and conditions to apply, compare the rates of fall of temperature of the two spheres. Indicate any further assumptions made in your calculation.

Worked example

46 *A copper calorimeter of mass* 50 g *containing* 100 cm^3 *of water cools* *at the rate of 2°C per minute when the temperature of the water is 50°C. If the water is replaced by 100 cm^3 of a liquid of specific heat capacity 2200 J kg^{-1} K^{-1} and relative density 0·8, what will be the rate of cooling when the temperature of the liquid is 40°C? Assume the cooling takes place according to Newton's law of cooling and that the constant temperature of the surroundings is 15°C in each case. (Specific heat capacity of water is 4200 J kg^{-1} K^{-1}; that of copper is 400 J kg^{-1} K^{-1}).*

According to Newton's law of cooling, the rate of loss of heat is proportional to the excess of temperature of the body above the surroundings. Now the rate of loss of heat from the calorimeter = heat capacity of calorimeter × rate of fall of temperature.
Hence in the first case,

$$(0{\cdot}05 \times 400 + 0{\cdot}1 \times 4200) \times 2 \propto (50-15) = k(50-15)$$

or
$$880 = k\,(50-15) \qquad (1)$$

where k is some constant determined by the nature and surface area of the calorimeter.

And in the second case,

$$(0{\cdot}05 \times 400 + {\cdot}1 \times {\cdot}8 \times 2200)x \propto (40-15) = k(40-15)$$

or
$$196x = k(40-15) \qquad (2)$$

where x is the required rate of fall of temperature in °C per minute.

We thus get, dividing equation (2) by equation (1),

$$\frac{196x}{880} = \frac{40-15}{50-15} = \frac{25}{35}$$

from which
$$x = \frac{880}{196} \times \frac{25}{35} = \underline{3{\cdot}21}\text{°C per minute}$$

47 State Newton's law of cooling. Under what conditions is it valid?

A calorimeter containing 40 g of water (specific heat capacity 4200 J kg^{-1} K^{-1}), cools from 60°C to 55°C in 1 min 36 sec. The same calorimeter, when containing 50 g of a liquid of specific heat capacity 2140 J kg^{-1} K^{-1} takes 1 min 8 sec to cool through the same temperature range under the same conditions of cooling. What is the heat capacity of the calorimeter?

48 Discuss the 'radiation correction' in calorimetry, indicating how it may be applied in a specific case.

A specimen of metal of mass 55 g was heated to a temperature of 100°C and subsequently quickly transferred to a calorimeter containing 100 g of water at a temperature of 11°C. The maximum temperature of 17·0°C was reached one minute after the transfer and after a further minute the temperature of the mixture had fallen to 16·5°C. If the heat capacity of the calorimeter was 42 J K^{-1}, what was the specific heat capacity of the metal? (Take the specific heat capacity of water as 4200 J kg^{-1} K^{-1}.)

49 Into a vessel containing 500 g of water at 10°C (the room temperature) is placed a 60-watt electric immersion heater. The temperature of the water is observed to rise to 20°C 7 minutes after switching on the heater. What is the highest temperature to which the water can be raised by the heater under the conditions of the experiment and what time elapses before this temperature is reached? Assume the heat losses conform to Newton's law throughout the temperature range involved. (Specific heat capacity of water = 4200 J kg^{-1} K^{-1}.)

50 The specific heat capacity of a liquid is known to vary with temperature: describe how you would determine the specific heat capacity of such a liquid *at a specific temperature.*

The temperature of a body falls from 40°C to 30°C in 5 minutes. What will be its temperature after a further 5 minutes? Assume the body to be loosing heat according to Newton's law of cooling, and that the constant temperature of the surroundings is 15°C.

Latent heats

51 Define *latent heat of vaporization* and describe an experiment to determine the specific latent heat of steam.

Steam at 100°C is blown into a vessel containing 1 kg of ice at −10°C. What mass of water at the temperature of its boiling point will the vessel eventually contain? Neglect heat losses and the heat capacity of the calorimeter. (Specific heat capacity of ice = $2{\cdot}1 \times 10^3$ J kg^{-1} K^{-1}. Specific latent heat capacities of ice and steam are $0{\cdot}336 \times 10^6$ and $2{\cdot}27 \times 10^6$ J kg^{-1} respectively.)

52 Describe Henning's method for determining the specific latent heat of vaporization of a liquid.

A heating coil immersed in a liquid takes a current of 2·5 ampere under a pressure of 20 volt. When at its boiling point 13·2 g of the liquid are evaporated in 5 minutes. Assuming 25 per cent of the heat generated by the coil is lost by radiation and convection effects, calculate a value for the specific latent heat of the liquid.

53 How have the specific heat capacities of the elements been obtained at low temperatures? What is the general nature of the results of such observations?

A piece of iron, of mass 0·35 g and at a temperature of 10°C is dropped into liquid oxygen at its boiling-point, viz.—183°C. The oxygen gas liberated is collected and found to have a volume of 82·5 cm^3 at 10°C and 75·2 cm of mercury pressure. If the specific latent heat of vaporization is $2{\cdot}14 \times 10^5$ J kg^{-1}, what is the mean specific heat capacity of iron between 10°C and −183°C?

54 Define *specific latent heat of fusion, melting point.*

When a quantity of molten lead at its melting-point is poured into a calorimeter containing oil, the temperature of the oil rises from 12·5°C to 25·0°C. The experiment is now repeated with the same mass of oil in the calorimeter, but the hot lead is not transferred until it has all just solidified. The temperature of the oil then rises from 12·5°C to 20·5°C. If the specific heat capacity of lead is 126 J kg^{-1} K^{-1}, what is its specific latent heat of fusion?

Worked example

55 *A mass of hot liquid of specific heat capacity* 1260 *J kg*$^{-1}$ *K*$^{-1}$ *is contained in a thin-walled vessel of negligible heat capacity and allowed to cool freely. The temperature-time curve is obtained and from it, it is found that the liquid cools at the rate of 2·9°C per minute just before solidification. The temperature then remains constant for* 40 *minutes and subsequently cooling proceeds at the rate of 3·1°C per minute immediately after complete solidification. What is the value of the specific latent heat of the substance and the specific heat capacity in the solid state?*

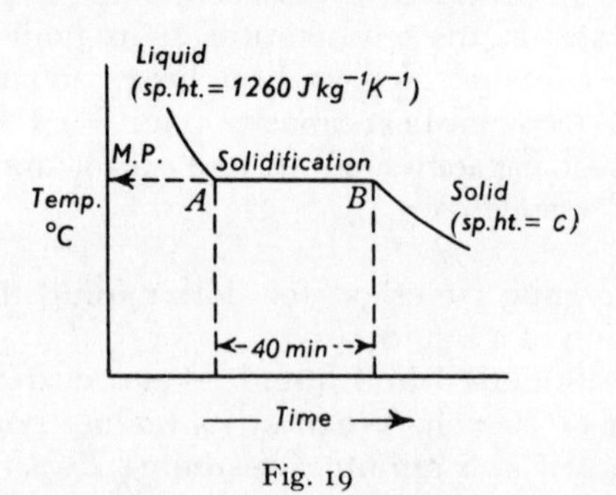

Fig. 19

The temperature-time curve for the substance is as indicated in the diagram. Let the mass of the cooling substance be *m* kg. Then, at A, the rate of loss of heat from the vessel

$= $ heat capacity of liquid substance $\times$ rate of fall of temperature at A
$= (m \times 1260) \times 2{\cdot}9 \quad \text{J min}^{-1}$.

At B, the rate of loss of heat from the vessel

$= $ heat capacity of solid substance $\times$ rate of fall of temperature at B
$= (m \times c) \times 3{\cdot}1 \quad \text{J min}^{-1}$.

But in each of these cases the rate of loss of heat is the same since the temperature is the same (melting-point of the substance) and the external conditions are the same.

Hence, $mc \times 3{\cdot}1 = m \times 1260 \times 2{\cdot}9$
i.e. c (specific heat capacity of solid substance) $= \underline{1179 \text{ J kg}^{-1}\text{K}^{-1}}$.

During the solidification stage AB the temperature remains constant since the rate of loss of heat from the vessel is just offset by the rate of release of latent heat as the substance changes state from liquid to solid. If l is the specific latent heat of fusion of the substance, then the rate of release of heat along AB

$$= \frac{ml}{40} \quad \text{J min}^{-1}$$

and the rate of loss of heat at this temperature (= rate of loss of heat at A)

$$= m \times 1260 \times 2{\cdot}9 \quad \text{J min}^{-1}$$

Hence $$\frac{ml}{40} = m \times 1260 \times 2{\cdot}9$$

or $$l = \underline{1{\cdot}462 \times 10^5} \quad \text{J kg}^{-1}$$

56 A vessel of thermal capacity 42 J K^{-1} containing 1 kilogramme of molten lead is allowed to cool under constant temperature conditions. It is found that the temperature of the lead falls from 333°C to 327°C in 32 seconds after which there is no further fall of temperature until the lapse of 15 min 4 sec. Explain these observations and estimate a value for the specific heat capacity of molten lead if the specific latent heat of fusion of lead is 21 840 J kg^{-1}.

57 What are the special advantages of latent heat calorimeters? Describe some form of latent heat calorimeter by means of which the specific heat capacities of gases can be found.

A piece of metal, of mass 20 g, is suspended in an enclosed vessel at a temperature of 15°C. When dry steam at 100°C is blown into the enclosure 0·31 g of steam condense on the metal. What is its specific heat capacity? (Specific latent heat of steam $= 2{\cdot}268 \times 10^6$ J kg^{-1}.)

58 Describe Bunsen's ice calorimeter and the method of using it to determine specific heat capacities.

Calculate the movement of the mercury thread along the capillary tube of a Bunsen's ice calorimeter when 3·25 g of a solid of specific heat capacity 235 J kg^{-1} K^{-1}, heated to a temperature of 100°C, is dropped into the apparatus. (Diameter of capillary tube = 0·4 mm, relative density of ice at 0°C = 0·92, specific latent heat of ice = 0·336 $\times 10^6$ J kg^{-1}.)

59 Explain the meaning and significance of the terms *internal latent heat, external latent heat.*

Use the data below to find the quantity of heat needed to do the internal work of converting 1 g of water at its normal boiling-point into steam at the same temperature.

Specific latent heat of steam = $2{\cdot}25 \times 10^6$ J kg^{-1}
Volume of 1 g of steam at 100°C = 1650 cm^3
Relative density of mercury = 13·6
$g = 9{\cdot}81$ m s^{-2}

Specific heat capacity of gases

60 Distinguish clearly between the two principal specific heat capacities of a gas and account for the difference in their numerical values.

Describe in detail how either of the principal specific heat capacities of a gas may be determined experimentally.

Worked example

61 *If the molar heat capacity of hydrogen at constant volume is* 20·2 *J* mol^{-1} K^{-1} *find the molar heat capacity at constant pressure and the speed of sound in hydrogen at 30°C using the following data:*

Molar volume of hydrogen at s.t.p. = $2{\cdot}24 \times 10^{-2}$ m^3 mol^{-1}
Density of mercury = $1{\cdot}36 \times 10^4$ *kg* m^{-3}
Acceleration of gravity = 9·81 *m* s^{-2}

The difference between the molar heat capacities of a gas at constant pressure (C_p) and constant volume (C_v) is given by the relation

$$C_p - C_v = R$$

where R is the molar gas constant and is obtained from the relation

$$R = \frac{pVm}{T}$$

V_m being the molar volume at s.t.p. $= 2{\cdot}24 \times 10^{-2}$ m^3 mol^{-1}, p being standard atmospheric pressure

$$= 0{\cdot}76 \times 1{\cdot}36 \times 10^4 \times 9{\cdot}81 \quad \text{N m}^{-2}$$

and T the standard temperature 273 K.

We thus have

$$\begin{aligned} C_p &= C_v + R \\ &= 20{\cdot}2 + \frac{0{\cdot}76 \times 1{\cdot}36 \times 10^4 \times 9{\cdot}81 \times 2{\cdot}24 \times 10^{-2}}{273} \\ &= 20{\cdot}2 + 8{\cdot}32 \\ &= \underline{28{\cdot}52}\ \text{J mol}^{-1}\ \text{K}^{-1} \end{aligned}$$

The speed of sound in a gas is given by the expression $c = \sqrt{\dfrac{\gamma p}{\rho}}$ where p and ρ are respectively the pressure and density of the gas and $\gamma = \dfrac{C_p}{C_v}$.

Now 1 mole of hydrogen gas has a mass of 2 g, hence its density at 0°C is

$$\frac{0{\cdot}002}{V_m} = \frac{0{\cdot}002}{2{\cdot}24 \times 10^{-2}}\ \text{kg m}^{-3}$$

and so, inserting the expression for the pressure from above, we get

$$\begin{aligned} c &= \sqrt{\frac{28{\cdot}52}{20{\cdot}2} \times \frac{0{\cdot}76 \times 1{\cdot}36 \times 10^4 \times 9{\cdot}81 \times 2{\cdot}24 \times 10^{-2}}{0{\cdot}002}} \\ &= 1266\ \text{m s}^{-1} \end{aligned}$$

But this is the value at 0°C. The speed of sound at 30°C is

$$\begin{aligned} & c_{0°\text{C}}\left(1 + \frac{30}{273}\right)^{\frac{1}{2}} \\ &= 1266\left(\frac{303}{273}\right)^{\frac{1}{2}} = \underline{1334}\ \text{m s}^{-1}. \end{aligned}$$

62 Distinguish between *isothermal* and *adiabatic* changes in a gas.

A monatomic gas, initially at atmospheric pressure, expands to four times its original volume (*a*) isothermally, (*b*) adiabatically. What is the final pressure of the gas in each case? The ratio of the specific heat capacity at constant pressure to that at constant volume for a monatomic gas $= \frac{5}{3}$.

63 Show that the ratio of the slope of an adiabatic curve to that of an isothermal curve at the point where they intersect is numerically equal to the ratio of the specific heats of the gas at constant pressure to that at constant volume.

A given mass of gas contained in a cylinder closed by a piston head has its volume halved by a sudden compression of the piston. If intially the pressure of the gas was 76 cm of mercury and its temperature was 27°C, find the pressure and temperature of the gas after the sudden compression. (The ratio of the specific heat capacity of the gas at constant pressure to that at constant volume may be taken as 1·4.)

Worked example

64 *The pressure of a mass of air at 0°C is suddenly halved. What is the resulting (a) volume, (b) temperature of the air? Ratio of specific heat capacities for air* = 1·41.

(*a*) The equation relating the pressure and volume in an adiabatic change is

$$pv^{\gamma} = \text{const.}$$

where γ is the ratio of the specific heat capacity at constant pressure to that at constant volume.

Hence if p_1, v_1 are the initial values of the pressure and volume, and p_2 v_2 the final values,

$$p_1 v_1^{\gamma} = p_2 v_2^{\gamma}$$

or
$$\left(\frac{v_2}{v_1}\right)^{\gamma} = \frac{p_1}{p_2}$$

Thus
$$\left(\frac{v_2}{v_1}\right)^{1\cdot 41} = 2 \quad \text{since } p_2 = \tfrac{1}{2}p_1$$

Takings logs,

$$1\cdot 41 \log\left(\frac{v_2}{v_1}\right) = \log 2 = 0\cdot 3010$$

or
$$\log\left(\frac{v_2}{v_1}\right) = \frac{0\cdot 3010}{1\cdot 41} = 0\cdot 2135$$

from which
$$\frac{v_2}{v_1} = 1\cdot 635$$

or, resulting volume $= \underline{1\cdot 635}$ times the original volume.

(*b*) The equation relating temperature and pressure in an adiabatic change is

$$\frac{T^{\gamma}}{p^{\gamma-1}} = \text{const.}$$

Hence, if T_1, p_1 are the initial values of the temperature and pressure, and T_2, p_2 the final values,

$$\frac{T_1^{\gamma}}{p_1^{\gamma-1}} = \frac{T_2^{\gamma}}{p_2^{\gamma-1}}$$

or

$$\left(\frac{T_1}{T_2}\right)^{\gamma} = \left(\frac{p_1}{p_2}\right)^{\gamma-1}$$

Inserting values,

$$\left(\frac{273}{T_2}\right)^{1\cdot41} = 2^{0\cdot41}$$

Taking logs,

$$1\cdot41\ (\log 273 - \log T_2) = 0\cdot41 \log 2 = 0\cdot1234$$

Hence $\quad \log 273 - \log T_2 = \dfrac{0\cdot1234}{1\cdot41} = 0\cdot0875$

giving $\quad \log T_2 = \log 273 - 0\cdot0875 = 2\cdot4362 - 0\cdot0875 = 2\cdot3487$

from which $\quad T_2 = 223\cdot2\text{ K}$ or $= -49\cdot8°\text{C}$

65 A given mass of air is enclosed in a vertical cylinder under a smoothly fitting piston. On being suddenly loaded, the piston attains a stationary position one-third the way down the cylinder. What will be the ultimate position of the piston if the ratio of the specific heats for air is 1·40?

66 Describe an acoustical method for determining the ratio of the two principal specific heat capacities of a gas.

Given that the density of hydrogen at s.t.p. is 9×10^{-2} kg m^{-3}, and that the velocity of sound in hydrogen at 0°C is 1260 m s^{-1}, calculate values for the two principal molar heat capacities of the gas. Assume values for any other physical constants required in your calculation.

67 The cylinder of an internal combustion engine has a volume of

1000 cm^3 and contains air mixed with a little petrol vapour, the mixture having a temperature of 17°C and being at atmospheric pressure. The mixture is now compressed adiabatically to one-fifth its volume after which it is fired (at constant volume) with the release of 420 J of heat energy. Use the data below to find (*a*) the temperature and pressure of the mixture at the end of the adiabatic compression, (*b*) the final temperature in the cylinder.

Specific heat capacity of the mixture at constant volume
$= 714 \text{ J kg}^{-1} \text{K}^{-1}$
Ratio of principal specific heat capacities of the mixture = 1·40
Density of the mixture = $1{\cdot}29 \text{ kg m}^{-3}$ at s.t.p.

68 Explain what is meant by the 'internal energy' of a substance.

1 g of hydrogen at s.t.p. has its volume halved by an adiabatic change. Calculate the change in the internal energy of the gas using the following data:

Molar gas constant = $8{\cdot}31 \text{ J mol}^{-1} \text{K}^{-1}$
Ratio of specific heats for hydrogen = 1·40.

69 Discuss the significance of (*a*) the difference, (b) the ratio of the specific heat capacities at constant pressure and volume of a gas.

Describe and give the theory of Clement and Desormes' method for determining the ratio of the specific heats of a gas. What are the chief sources of error in this experiment? How have they been overcome?

Unsaturated vapours and vapour pressure

70 Compare the properties of saturated and unsaturated vapours.

A sample of moist air at 12°C and at a pressure of 76 cm of mercury is *just* saturated with water vapour. Calculate the pressure of the moist air if the volume is (*a*) halved, (*b*) doubled, under isothermal conditions. (Saturated vapour pressure of water at 12°C = 10·5 mm of mercury.)

71 The space above the mercury in a mercurial barometer is vitiated with water vapour (saturated) and traces of air, the pressure of the former being 12 mm of mercury. If the mercury in the tube stands at a height of 72 cm above the mercury in the trough, at what height will it stand on depressing the tube to the extent of reducing the space above the mercury in the tube to one-third of its original volume? The true barometric height is 76 cm of mercury throughout.

72 Give the details of a method of determining the saturated vapour pressure of water at temperatures between 50°C and 100°C.

100 cm^3 of air are saturated with water vapour at 100°C, the pressure of the mixture being 100 cm of mercury. What increase of pressure is needed to reduce the volume to one-quarter of its original volume if, at the same time, the mixture is cooled to a temperature of 75°C? (Saturated vapour pressure of water at 75°C = 28·9 cm of mercury.)

73 How does the process of *boiling* differ from that of *evaporation*? A closed vessel of fixed volume contains air which is just saturated with water vapour at 100°C. On raising the temperature to 150°C the pressure inside the vessel becomes 2 atmospheres. Find what would be the pressure inside the vessel if the temperature were to be reduced to 0°C. State the assumptions made in arriving at your answer.

Worked example

74 *A horizontal tube of uniform bore, closed at one end, has some air entrapped by a small quantity of water. If the length of the enclosed air column is* 20 *cm at* 12°*C, what will it be if the temperature is raised to* 38°*C, the atmospheric pressure remaining constant at* 75·0 *cm of mercury throughout? (s.v.p. of water vapour: at* 12°*C* = 10·5 *mm, at* 38°*C* = 49·5 *mm of mercury.)*

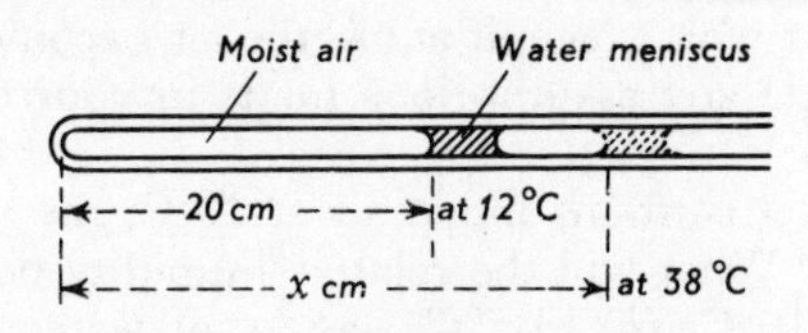

Fig. 20

Under all conditions of the tube the meniscus will adjust itself to a position of equal pressure on both sides of it. Hence the total pressure of the moist air entrapped in the tube = atmospheric = 75·0 cm of mercury.

At 12°C (285 K = T_1)

Pressure of water vapour = 10·5 mm

∴ Partial pressure (p_1) of the air at this temperature

$= 750 - 10{\cdot}5$

$= 739{\cdot}5$ mm

and the volume (v_1) of this air = 20×2 cm^3 where $\alpha(cm^2)$ is the cross-sectional area of the tube.

At 38°C (311 K = T_2)

Pressure of water vapour = 49·5 mm

$\therefore$ Partial pressure (p_2) of the air at this temperature

$$= 750 - 49{\cdot}5$$
$$= 700{\cdot}5 \text{ mm}$$

and the new volume (v_2) of the air $= x \times \alpha$ cm^3 (see figure).

Hence, applying the gas laws to the dry air component of the entrapped moist air, we have

$$\frac{p_1 v_1}{T_1} = \frac{p_2 v_2}{T_2}$$

i.e.
$$\frac{739{\cdot}5 \times 20\alpha}{285} = \frac{700{\cdot}5 \times x\alpha}{311}$$

giving
$$x = \frac{739{\cdot}5}{700{\cdot}5} \times \frac{311}{285} \times 20$$
$$= \underline{23{\cdot}04 \text{ cm}}$$

75 The closed end of a Boyle's law apparatus contains moist air which is initially unsaturated and exerts a pressure of 40·5 cm of mercury. When slowly compressed to half its original volume the pressure becomes 80·4 cm of mercury and, after a further slow compression to one-quarter of the original volume, the pressure becomes 159·6 cm of mercury. Calculate the original pressure of the water vapour and state at what stage, if at all, the air becomes saturated with water vapour. State any assumptions made in your calculation.

76 A closed vessel contains moist air at 20°C, the relative humidity being 40 per cent. What will the relative humidity become on cooling the vessel (*a*) to 10°C, (*b*) to 5°C? (s.v.p. of water vapour: at 20°C = 17·51 mm, at 10°C = 9·21 mm, at 5°C = 6·54 mm of mercury.)

77 Describe a method of determining the relative humidity of the air.

On a day when the air temperature is 16°C, the relative humidity of the air is found to be 75 per cent. What percentage of the mass of the water vapour present will be deposited if the air temperature were to fall suddenly to 4°C? (s.v.p. of water vapour: at 4°C = 6·1 mm, at 16°C = 13·6 mm of mercury.)

78 Define *dewpoint* and *relative humidity* and obtain a relation between these two quantities.

Calculate the mass of 1000 cm^3 of moist air at 20°C and 770 mm pressure using the following data:

Density of dry air at s.t.p. = 1·293 kg m^{-3}.

Density of water vapour is five-eighths that of dry air under the same conditions.

s.v.p. of water vapour at 20°C = 17·5 mm of mercury.

79 Derive an expression for the change in the vapour pressure of a liquid occasioned by the curvature of the liquid surface with which is in contact. Discuss the significance of this change in meteorological phenomena.

Calculate the saturated vapour pressure over the surface of a water droplet of diameter 0·001 mm at 12°C, the pressure of the air being 760 mm at the time.

Surface tension of water at 12°C = $7{\cdot}2 \times 10^{-2}$ N m^{-1}.

Saturated vapour pressure of water at 12°C = 10·5 mm of mercury.

Density of dry air at s.t.p. = 1·293 kg m^{-3}.

Density of water vapour is five-eighths that of dry air under the same conditions.

80 What is an *isothermal*?

Describe the work of Andrews in obtaining the isothermals of carbon dioxide and discuss the importance of these researches in extending the knowledge of the conditions for gas liquefaction.

81 Explain the terms *critical temperature* and *inversion temperature* as used in the liquefaction of gases.

Describe, in outline, stressing the physical principles involved, a method for obtaining continuous supplies of liquid air.

82 Outline the different methods used to obtain low temperatures and discuss in particular the method used to liquefy such gases as hydrogen and helium.

Kinetic theory of gases

83 Give a general account of the kinetic theory of matter indicating how it explains the essential differences between the solid, liquid and gaseous states. Discuss in qualitative terms how the theory accounts for (*a*) the phenomenon of surface tension in liquids, (*b*) diffusion of gases.

84 Derive an expression for the pressure of an ideal gas in terms of its density and the mean square velocity of its molecules. State the assumptions made and indicate the modifications needed to account for the behaviour of real gases.

85 Use the following data to find the number of molecules in 1 cm^3 of a gas at (*a*) 0°C and 10^{-4} mm mercury pressure, (*b*) at 50°C and 10^{-4} mm mercury pressure:

Avogadro constant = $6{\cdot}02 \times 10^{23}$ mol^{-1}
Molar volume at s.t.p. = $2{\cdot}24 \times 10^{-2}$ m^3 mol^{-1}
Relative density of mercury = 13·6
Acceleration of gravity = 9·81 m s^{-2}

86 Discuss the kinetic theory interpretation of temperature as embodied in the simple kinetic theory of gases.

At what temperature will oxygen molecules have the same mean velocity as the molecules of hydrogen gas have at −50°C?

Worked example

87 *Given that the molar gas constant (R) has a value of 8·31 J mol^{-1} K^{-1}, calculate the following root mean square molecular velocities: (a) hydrogen gas at 0°C, (b) hydrogen gas at 50°C, (c) oxygen gas at 0°C. (Assume conditions of standard pressure throughout.)*

From the elementary kinetic theory of gases, the pressure (p) of a gas of density ρ is given by the expression

$$p = \tfrac{1}{3}\rho c^2$$

where c is the root mean square velocity of the gas molecules.

Now $$\rho = \frac{\text{mass } (m)}{\text{volume } (v)}$$

So $$pv = \tfrac{1}{3} mc^2 = RT$$

Hence we see that $$c = \sqrt{\frac{3pv}{m}} = \sqrt{\frac{3RT}{m}}$$

(*a*) For hydrogen gas 1 mole has a mass of 0·002 kg

$$\therefore c_{\text{hydrogen}} \text{ at } 0°\text{C} = \sqrt{\frac{3 \times 8{\cdot}31 \times 273}{0{\cdot}002}}$$

$$= \underline{1{\cdot}845 \times 10^3 \text{ m s}^{-1}}$$

(*b*) Now for a given gas (m const.) we have, from the above equation $c \propto \sqrt{T}$, hence for hydrogen gas

$$\frac{c_{50°C}}{c_{0°C}} = \sqrt{\frac{323}{273}}$$

or
$$c_{50°C} = c_{0°C}\sqrt{\frac{323}{273}} = 1{\cdot}845 \times 10^3 \sqrt{\frac{323}{273}}$$
$$= \underline{2{\cdot}007 \times 10^3 \text{ m s}^{-1}}$$

(*c*) Also from the above expression, for different gases at the same temperature, $c \propto \dfrac{1}{m}$

Hence
$$\frac{c_{\text{oxygen}}}{c_{\text{hydrogen}}} = \sqrt{\frac{m_{\text{hydrogen}}}{m_{\text{oxygen}}}} = \sqrt{\frac{0{\cdot}002}{0{\cdot}032}} = \frac{1}{4}$$

Thus c_{oxygen} at 0°C $= \dfrac{c_{\text{hydrogen}} \text{ at } 0°C}{4}$

$$= \frac{1{\cdot}845 \times 10^3}{4} = \underline{4{\cdot}61 \times 10^2 \text{ m s}^{-1}}$$

88 Obtain an expression for the mean molecular velocity of a gas in terms of its pressure and density.

Calculate the value of this mean velocity for the molecules of oxygen gas at a temperature of 27°C. Comment on the nature of the mean value so obtained. (Molar volume for gases at s.t.p. $= 2{\cdot}24 \times 10^{-2}$ m^3 mol^{-1}.)

89 Give an account of the Brownian movement and indicate what molecular information can be obtained from observation of the effect in colloidal suspensions.

90 State clearly what you understand by (a) *root mean square velocity,* (b) *mean free path* as applied to a gas.

Calculate the root mean square velocity of oxygen molecules at 0°C from the following data:

Velocity of sound in oxygen gas at 0°C = 317 m s^{-1}.

Specific heat capacity of oxygen gas at constant pressure
$= 815{\cdot}6$ J kg^{-1} K^{-1}.

Molar gas constant $= 8{\cdot}31$ J mol^{-1} K^{-1}.

Conductivity

91 Define *coefficient of thermal conductivity*.

In a steel boiler the plates are 8 mm thick and 30 kg of water are evaporated per minute per square metre of boiler surface. Calculate the temperature drop across the boiler plates if the thermal conductivity of steel is 46·2 W m^{-1} K^{-1} and the specific latent heat of steam is $2{\cdot}25 \times 10^6$ J kg^{-1}.

92 Describe a method of obtaining the thermal conductivity of a good conductor.

A solid lump of ice at 0°C fully occupies the space inside a cubical metal box of side 10 cm. If the metal has a uniform thickness (sides, top and bottom) of 2 mm, and its thermal conductivity is 42 W m^{-1} K^{-1}, find how long it will take for all the ice to melt when the box is completely immersed in water maintained at 100°C. (Density of ice = $9{\cdot}2 \times 10^2$ kg m^{-3}, specific latent heat of fusion of ice = $3{\cdot}36 \times 10^5$ J kg^{-1}.)

93 Describe a method of determining the thermal conductivity of a poor conductor available in the form of a thin sheet.

A room is heated by a 2-kilowatt electric radiator. Assuming that the only heat leakage is by conduction through the windows of total area 2 m^2 and thickness 3 mm, calculate the steady room temperature when the temperature of the air outside is 5°C. (Thermal conductivity of glass = 1·26 W m^{-1} K^{-1}.)

94 A copper sphere of mass 100 g is suspended by a piece of copper wire, of length 20 cm and diameter 1 mm, the upper end of which projects into a steam jacket which can be maintained at 100°C. If the temperature of the apparatus is 10°C, find the initial rate of rise of temperature of the copper sphere after the steam is turned on in the steam-jacket. Thermal conductivity of copper = 386·4 W m^{-1} K^{-1}, specific heat capacity of copper = 380 J kg^{-1} K^{-1}. (Assume no lateral losses of heat from the copper wire).

95 Discuss the transfer of heat along an exposed metal bar which is heated at one end.

One end of a long uniform metal bar of diameter 1 cm is maintained at a constant high temperature and in the final steady state it is found that the temperature gradients at two points A and B along the bar are 3·5°C per cm and 1·5°C per cm respectively. If A and B are 50 cm apart and the thermal conductivity of the metal is 252 W m^{-1} K^{-1}, calculate the average rate of loss of heat per cm^2 of surface of the bar between A and B.

96 Steam at 100°C is supplied through a thin copper delivery pipe 8 cm in diameter wrapped with a 1 cm thickness of asbestos lagging of thermal conductivity 0·126 W m^{-1} K^{-1}. If the temperature of the surrounding air is 10°C, calculate the approximate amount of steam condensed per metre run of the pipe per minute. (Specific latent heat of steam = $2{\cdot}27 \times 10^6$ J kg^{-1}.)

Worked example

97 *A composite slab is made of two parallel layers of differently conducting materials, A and B, in close contact. The thermal conductivities of A and B are 70 and 200 W m^{-1} K^{-1}, and their thicknesses are 4·5 and 2·5 cm respectively. If the outer surface of A is maintained at 100°C and the outer surface of B is maintained at 0°C, find the temperature of the common surface of the two materials and the heat conducted per cm^2 through the composite slab in one minute.*

Let the temperature of the common surface in the steady state = θ°C. In this steady state the quantity of heat (Q) conducted per second across unit area of B will be the same as that conducted across A,

i.e.
$$Q = k_1\left(\frac{100-\theta}{d_1}\right) = k_2\left(\frac{\theta-0}{d_2}\right)$$

or
$$70\left(\frac{100-\theta}{0{\cdot}045}\right) = 200\left(\frac{\theta-0}{0{\cdot}025}\right)$$

which, on re-arranging, becomes

$$\frac{7}{36}(100-\theta) = \theta$$

giving
$$\underline{\theta = 16{\cdot}28°\text{C}}$$

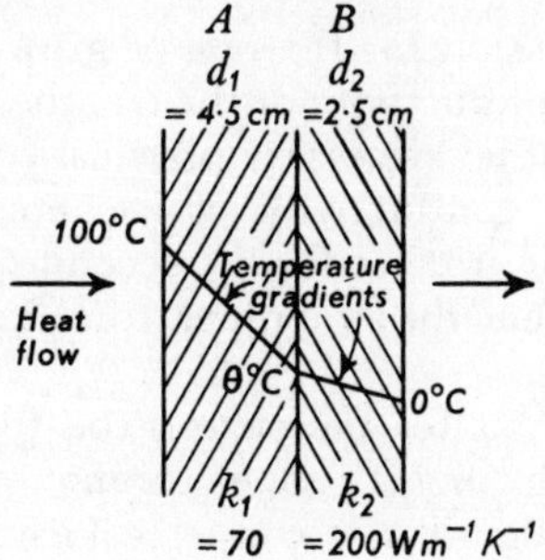

Fig. 21

The quantity of heat conducted per minute across an area of 1 cm^2 of A (and hence across the composite slab)

$$= 70 \times \frac{1}{10^4} \frac{(100-16\cdot28)}{0\cdot045} \times 60$$

$$= \underline{781\cdot4 \text{ J}}$$

98 The metal plates of a boiler are 1 cm thick and conduct sufficient heat to evaporate 50 kg of water per square metre of plate area per hour. What is the temperature difference across the plates? If the inside of the boiler becomes encrusted with 'boiler scale' to a thickness of 5 mm, what temperature difference must now be maintained across the total thickness of deposit and plate to maintain the same rate of heat flow? (Specific latent heat of vaporization = $2\cdot27 \times 10^6$ J kg^{-1}. Thermal conductivities of metal plate and scale deposit are 420 and 4·2 W m^{-1} K^{-1}.)

99 The faces of an iron plate 2 cm thick are covered by water films 0·1 mm thick. If the overall temperature drop across films and plate is 60°C, calculate (*a*) the temperature drop across the plate itself and (*b*) the quantity of heat conducted through each square metre of surface of the plate per minute. (Thermal conductivities of iron and water are 75·6 and 0·63 W m^{-1} K^{-1} respectively.)

100 A uniform layer of ice, 5 cm thick, has already formed on a pond on a day when the temperature of the air immediately above the surface is −7·5°C. Calculate approximately how long it will take for the ice layer to increase in thckness by 1 mm given that the thermal conductivity of ice is 2·1 W m^{-1} K^{-1}, its density is 920 kg m^{-3} and its specific latent heat is $3\cdot36 \times 10^5$ J kg^{-1}.

101 Obtain an expression for the rate of growth of an ice layer on a pond in terms of the conductivity of the ice, the air temperature above the ice layer and the other necessary physical constants involved.

Using the data for conductivity of ice, etc., given in the above problem, find how long it takes for the second centimetre thickness of ice to form on a pond when the air temperature above the layer is −10°C.

102 Describe a method for measuring the thermal conductivity of **either** (*a*) glass tubing, **or** (*b*) rubber tubing.

A glass U-tube is placed in a copper calorimeter (of heat capacity 105 J K^{-1}) containing 1 kg of water at 10°C. If the internal and external radii of the tube are respectively 2·5 and 2·7 cm, and the total length of the tube submerged below the water is 30 cm, calculate the initial rate of rise of temperature of the water in the calorimeter on passing steam at 100°C through the tube. (Thermal conductivity of glass = 1·26 W m^{-1} K^{-1}. Ignore heat losses from the calorimeter and assume steady state conditions.)

Worked example

103 *Steam at 100°C is conveyed through an iron steam pipe having internal and external radii of 3 and 4 cm respectively. Calculate (i) the external temperature of the pipe and (ii) the mass of steam condensed per hour per metre of the pipe when the surrounding air temperature is 10°C if the thermal conductivity of the iron is 63 W m^{-1} K^{-1} and the emissivity of the surface of the pipe = 25·2 J m^{-2} per °C excess temperature. (Specific latent heat of steam = $2{\cdot}27 \times 10^6$ J kg^{-1}.)*

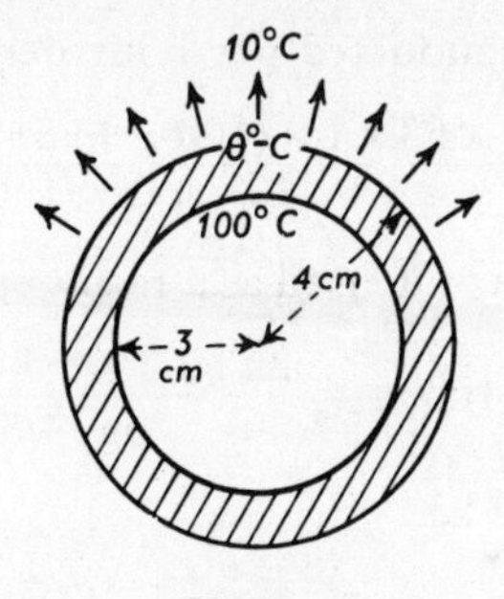

Fig. 22

The quantity (Q) of heat conducted per second radially across length l of the pipe

$$= \frac{2\pi kl(\theta_1 - \theta_2)}{\log_e \frac{r_2}{r_1}}$$

Where k is the conductivity of the material, θ_1 and θ_2 the inside and outside temperatures, and r_1 and r_2 the inside and outside radii. Hence, if θ is the temperature of the external surface we have

$$Q = \frac{2\pi \times 63 \times l(100 - \theta)}{\log_e 4/3}$$

The heat thus conducted through the pipe is lost by radiation from its outside surface. This heat loss, per second, for a length l of the pipe

$$= \text{emissivity} \times \text{surface area} \times \text{temperature excess}$$

$$= 25{\cdot}2 \times 2\pi \frac{4}{100} l \times (\theta - 10)$$

Hence

$$\frac{2\pi \times 63 l(100-\theta)}{\log_e 4/3} = 25{\cdot}2 \times 2\pi \frac{4}{100} l(\theta - 10)$$

or, rearranging,

$$100 - \theta = \log_e \frac{4}{3} \times \frac{25{\cdot}2 \times 4}{63 \times 100} (\theta - 10)$$

from which $\theta = \underline{99{\cdot}5^\circ\text{C}}$

The quantity of heat conducted per hour per metre length of pipe

$$= \frac{2\pi \times 63 \times 1 \times (100 - 99{\cdot}5) \times 3600}{\log_e 4/3}$$

If a mass m of steam is condensed as a result of this heat loss, then

$$2{\cdot}27 \times 10^6\, m = \frac{2\pi \times 63 \times 0{\cdot}5 \times 3600}{\log_e 4/3}$$

from which $m = \underline{1{\cdot}09 \text{ kg}}$

104 Obtain an expression for the rate of radial flow of heat per centimetre length of a hollow cylindrical conductor of internal and external radii r_1 and r_2 respectively if the corresponding surface temperatures in the steady state are θ_1 and θ_2.

A wire of diameter 1 mm carries a current of 5 ampere and is covered uniformly with a cylindrical layer of insulating material which has a coefficient of thermal conductivity of $0{\cdot}126 \text{ W m}^{-1} \text{ K}^{-1}$ and a diameter of 1 cm. If the resistivity of the wire is $1{\cdot}8 \times 10^{-8}$ ohm-metre units, what is the temperature difference between the inner and outer surfaces of the insulating covering?

105 The annular space between two thin concentric copper shells of radii 5 and 10 cm respectively is filled with a substance whose thermal conductivity is to be measured. An electric heater rated at 15 watt is enclosed in the inner sphere and when steady state conditions have been attained a temperature difference of 75°C is observed between the two copper shells. Calculate the thermal conductivity of the material from this data giving the theory of your method.

Radiation

106 Give a short account of the properties of radiant heat indicating the extent of agreement between these properties and those of light. Give the experimental evidence in support of your argument.

107 Describe and explain the mode of action of some instrument suitable to detect radiant heat. How would you use the instrument in (*a*) comparing the emissive powers of two surfaces, (*b*) determining the diathermanency of a sheet of glass?

108 Give an account of Prevost's theory of exchanges and discuss the interpretation of the temperature of a body implicit in this theory.

109 What is meant by 'black body radiation'? How is such radiation realized in practice? Give an account of the distribution of black body radiation between the wavelengths and its variation with temperature.

110 State Stefan's law of radiation and indicate the conditions of its application.

A solid metal sphere is found to cool at the rate of 1·2°C per minute when its temperature is 127°C. At what rate will a sphere of three times the radius cool when at a temperature of 327°C if, in each case, the external temperature is 27°C. Assume Stefan's law to apply in each case.

111 Give a critical discussion of Fery's total radiation pyrometer. How is such an instrument calibrated and what modifications are necessary to the standard instrument to enable very high temperatures (*c.* 2000°C) to be measured?

112 What do you understand by the terms *solar constant, Stefan's constant*?

If the values of the above constants are respectively $1{\cdot}35 \times 10^3$ W m^{-2} and $5{\cdot}75 \times 10^{-8}$ W m^{-2} K^{-4}, estimate a value for the surface temperature of the sun, given that the radius of the sun is $6{\cdot}96 \times 10^5$ km and that its mean distance from the earth is $1{\cdot}49 \times 10^8$ km.

LIGHT

Reflection at plane surfaces

1 Two plane mirrors are arranged so that their reflecting surfaces make an angle of θ with each other. Prove that a ray of light incident at any angle on one mirror and subsequently reflected from the other mirror emerges at a constant deviation of 2θ with its original direction.

2 Account for the multiple images formed by two facing parallel mirrors.

After five reflections between two parallel mirrors an image of an object placed 1 cm in front of one of the mirrors is located at a distance of 17 cm behind it. What is the distance between the mirrors?

3 Prove that when a plane mirror is rotated through a given angle, a ray reflected from it is rotated through twice that angle.

Describe, with illustrative diagrams, **two** instruments which make use of this principle.

Reflection at curved surfaces

4 A point object placed in front of a concave spherical mirror produces a fourfold magnified real image. On moving the object 10 cm towards the mirror a similarly magnified virtual image is formed. Find the focal length of the mirror.

5 Describe **two** methods of finding the focal length of a convex spherical mirror.

Calculate the nature and position of the image of an object 10 cm distant from a convex mirror of focal length 20 cm when the object is (*a*) real, (*b*) virtual.

6 A pin is mounted vertically in front of a convex mirror at a distance of 40 cm from it. A plane mirror is now placed between the convex mirror and the pin at such a height that an eye looking towards the mirrors sees images of the top half of the pin in the convex mirror and of the lower half of the pin in the plane mirror. Non-parallax coincidence between these two images is obtained when the plane mirror is adjusted to a position 16 cm in front of the convex mirror. Calculate the focal length of the convex mirror.

Worked example

7 *An observer finds that 20 cm is the least distance from his eye at which he can place a convex mirror of focal length 15 cm in order to see his own eye clearly. What is his least distance of distinct vision?*

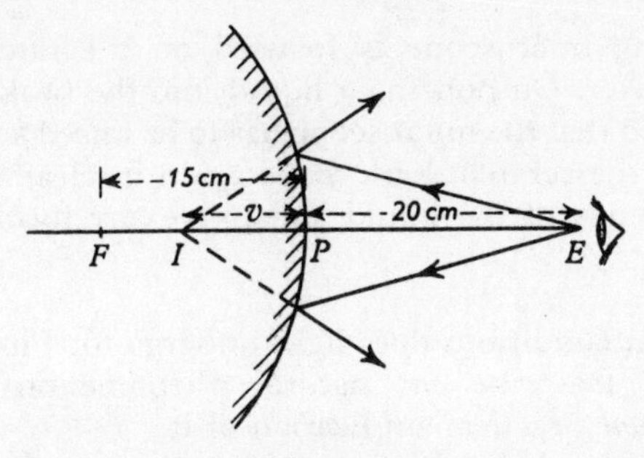

Fig. 23

The eye sees itself by means of the diverging rays reflected from the mirror. They appear to originate at I (see diagram) and hence the least distance of distinct vision of the eye $IE = 20+v$.

Now using the mirror formula $\frac{1}{u}+\frac{1}{v} = \frac{1}{f}$, we have

$$\frac{1}{20}+\frac{1}{v} = \frac{1}{-15}$$

(negative sign for f since a convex mirror has a virtual principal focus)

from which $v = \underline{-8\frac{4}{7}}$ cm.

Hence, disregarding the negative sign (which merely indicates that the image is virtual), we see that the required least distance of distinct vision

$$= 20+8\tfrac{4}{7} = \underline{28\tfrac{4}{7}\text{ cm}}$$

8 A small convex mirror has its reflecting surface facing towards that of a wide-aperture concave mirror which is mounted coaxially with the convex mirror. If the focal length of each mirror is 1 m, find what must be the distance between them for a real image to be formed at the surface of the concave mirror when rays from a distant object fall on it.

9 A convex lens of focal length 10 cm is placed 20 cm in front of the reflecting surface of a convex spherical mirror. It is then found that an object placed on the side of the lens remote from the mirror, coincides with its own image when 12·5 cm from the lens. What is the focal length of the convex mirror?

Refraction at plane surfaces. Prisms

10 A travelling microscope is focused on a scratch mark on the bottom of a beaker. On pouring a liquid into the beaker to a depth of 6·0 cm it is found that the microscope has to be raised a vertical distance of 1·52 cm for the scratch mark to be again in clear focus. Calculate the refractive index of the liquid and prove any formula used in the calculation.

11 Under what conditions does light undergo total internal reflection in a medium? Describe *one* natural phenomenon involving total reflection, and *one* practical application of it.

When a parallel-sided slab of glass 3 cm thick is placed over a small illuminated aperture in a white screen, it is observed that the illumination of the screen is increased at all points outside a circle centred at the aperture and of radius 5·6 cm. Explain this and calculate a value for the refractive index of the glass.

12 Describe and explain a total reflection method for determining the refractive index of a liquid.

If the refractive indices from air to glass and from air to water are 1·50 and 1·33 respectively, calculate a value for the critical angle for a water-glass surface.

13 A few drops of a liquid are squashed into a thin film below the base of a glass cube one vertical face of which is illuminated with sodium light. An eye looking into the cube through the opposite vertical face notices that the light ceases to be internally reflected from the base when the position of the eye is such that the emergent ray makes an angle with the normal greater than 48°. If the refractive index of the glass is 1·52, calculate the refractive index of the liquid for sodium light.

14 A concave mirror is placed on a table so that its optical axis is vertical. A pin held above the mirror coincides with its own image when at a distance of 20 cm from the mirror. A thin layer of liquid is now poured into the mirror when it is found that the pin has to be moved 5·2 cm for it to be again in coincidence with its image. Explain the formation of these images and calculate the refractive index of the liquid.

15 Prove that the lateral displacement, d, of a ray of light incident at an angle i on a parallel-sided slab of glass of thickness t and refractive index n is given by

$$d = t \sin i \left(1 - \frac{\cos i}{\sqrt{n^2 - \sin^2 i}}\right)$$

Suggest an experimental method by means of which a value of n may be measured from lateral displacement measurements.

16 Two immiscible liquids lie one on top of the other in a tank in equal layers each of depth 10 cm. If the lower liquid has a refractive index of 1·65 and the upper liquid a refractive index of 1·33, what is the apparent depth of the liquids when viewed vertically from above?

17 Assuming the refractive index of the air to decrease at the rate of 0·01 per cent for every metre of vertical ascent, calculate the height at which a ray, initially travelling at ground level at 45° to the surface, will undergo total internal reflection.

18 The refractive index of a column of liquid 10 cm deep increases uniformly from a value 1·33 at the surface to a value 1·63 at the bottom. What is the apparent depth of the liquid column when viewed vertically from above?

19 Establish the relation connecting the refractive index (n) of a prism, the refracting angle (A) and the minimum deviation (D) of a ray passing through it.

What is the minimum deviation for a prism of refractive index 1·5 and having a refracting angle of 60°C? Find also the angle of incidence of the ray undergoing minimum deviation with this prism.

20 A 60° prism has a refractive index of 1·52. Calculate the angle of incidence of the ray which on being refracted into the prism *just* undergoes total internal reflection at the opposite face.

21 Describe how you would use a spectrometer to determine a value of the refractive index of a glass prism.

What is the largest angle of a glass prism of refractive index 1·5 if a ray of light is to be refracted through the prism?

Worked example

22 *A parallel beam of monochromatic light incident nearly parallel to one face of a triangular prism enters the prism and after refraction at a second face emerges in a direction perpendicular to its first face. If the refractive index of the material of the prism for the light is* 1·50, *determine the refracting angle of the prism.*

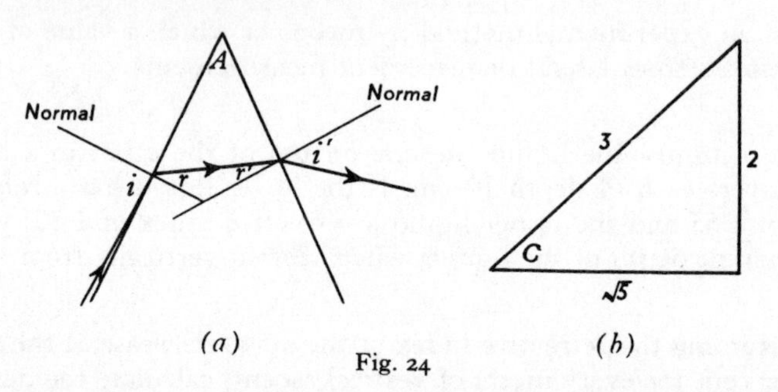

Fig. 24

The diagram shows the path of a ray through the prism of refracting angle A. The deviation of the ray is $(i-r)$ at the first face and $(i'-r')$ at the second face.

Hence the total deviation D

$$= (i-r)+(i'-r') = (i+i')-(r+r') = (i+i')-A$$

(since $r+r' = A$).

Now this deviation is 90°, and taking i as 90°, we have

$$90 = (90+i')-A \quad \text{or} \quad i' = A.$$

If i is 90°, r must be equal to the critical angle (c) of the glass and hence $r' = (A-c)$. Now applying Snell's law

$$\sin i' = n \sin r'$$

i.e.

$$\sin A = \tfrac{3}{2} \sin (A-c)$$

$$= \tfrac{3}{2} (\sin A \cos c - \cos A \sin c)$$

Dividing this by $\sin A$, we have

$$1 = \tfrac{3}{2} (\cos c - \cot A \sin c)$$

Now $c = \sin^{-1} \dfrac{1}{n} = \sin^{-1} \left(\dfrac{1}{1\cdot 5}\right)$

Hence $\sin c = \dfrac{2}{3}$ and $\cos c = \dfrac{\sqrt{5}}{3}$ (see diagram).

Inserting these values in the above equation, we have

$$1 = \frac{3}{2}\left(\frac{\sqrt{5}}{3} - \frac{2}{3}\cot A\right)$$

from which $$\cot A = \frac{\sqrt{5}-2}{2} = 0{\cdot}118$$

giving $$A = \underline{83^\circ 16'}.$$

23 Two rays, whose angles of incidence differ by 10°, undergo the same deviation of 50° on being refracted through a prism of refracting angle 60°. What is the refractive index of the prism?

24 A ray is refracted through a 60° prism in such a way that the angle of emergence is twice the angle of incidence. If the ray undergoes a deviation of 30°, calculate the refractive index of the material of the prism.

25 What is the refractive index of a 60° prism if a ray entering at grazing incidence on one face emerges at an angle of 45° with the adjoining face?

26 The faces of a glass wedge of angle 5° are partially silvered and a ray of light incident normally on one of the faces emerges after two internal reflections at an angle of 18° with its original direction. Calculate the refractive index of the wedge. What is the largest number of internal reflections that the above ray can undergo if the light is finally to emerge from the wedge?

27 Derive a formula for the deviation of a ray making a small angle of incidence on a transparent prism of refractive index n and with a small refracting angle A. Hence, or otherwise, deduce the formula for the focal length of a thin lens in terms of the radii of its surfaces and the refractive index of the lens material.

Refraction at curved surfaces. Lenses

28 A convex lens mounted in the roof of a darkened hut throws, on a horizontal table 3 m below the lens, an image of an aeroplane in horizontal flight at 6000 m. The image of the aeroplane is observed to move across the table at the rate of 8 cm s^{-1}. What is the speed of the aeroplane in km hr^{-1}?

29 Give the experimental details of **two** methods of finding the focal length of a converging lens.

Such a lens is placed on an optical bench and is moved about until it gives a real image of an illuminated object at a minimum distance from the latter. This distance is found to be 60·0 cm. What is the focal length of the lens?

30 Show that, in general, there are two positions which a convex lens can occupy when producing a real image on a screen of an illuminated object a fixed distance away.

If the ratio of the image sizes so produced is 9, and the distance between the two positions of the lens is 20 cm, calculate the focal length of the lens and the distance between the object and screen.

31 Describe the 'displacement method' of finding the focal length of a converging lens. What are the special merits of this method?

A converging lens is mounted in an inaccessible position inside a tube which is supported with its axis horizontal between an illuminated aperture and a screen 1 m away. It is found that clearly focused images of the aperture are formed on the screen when the end of the tube nearer the screen is 53 cm from it, and again when 33 cm from it. Find the focal length of the lens and its position in the tube.

Worked example

32 *A small linear object is placed perpendicular to the axis of a thin converging lens and a sharp image is produced on a screen. The magnification is* 1·59. *The screen is moved a distance* 7·6 *cm towards the lens and the position of the object is adjusted so that a sharp image is again produced on the screen. The magnification is now* 1·20. *Determine the focal length of the lens.*

Now from the lens formula $\frac{1}{u}+\frac{1}{v} = \frac{1}{f}$ we have,

$$m = \frac{v}{u} = \frac{v}{f} - 1$$

Hence for 1st position,

$$1{\cdot}59 = \frac{v_1}{f} - 1 \quad \text{or} \quad v_1 = 2{\cdot}59f$$

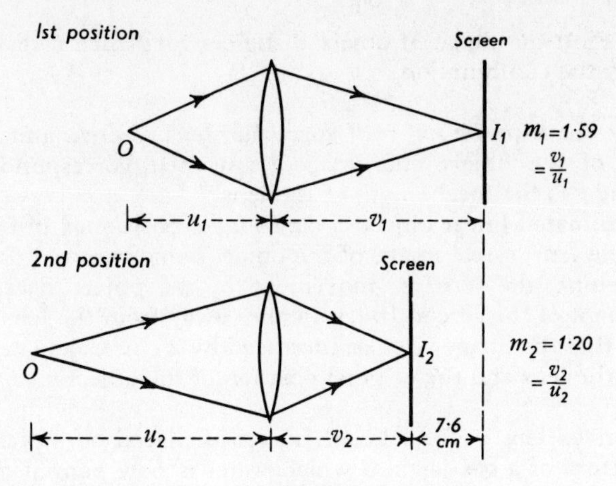

Fig. 25

and for 2nd position

$$1{\cdot}20 = \frac{v_2}{f} - 1 \quad \text{or} \quad v_2 = 2{\cdot}20f$$

But $\quad v_2 = v_1 - 7{\cdot}6$

Accordingly $\quad 2{\cdot}20f = 2{\cdot}59f - 7{\cdot}6$

from which $\quad f = \dfrac{7{\cdot}6}{0{\cdot}39} = \underline{19{\cdot}5\text{ cm}}$

33 An eye positioned 15 cm from a convex lens sees an image of itself by parallel rays when looking through the lens towards a plane mirror placed 20 cm behind the lens. Calculate possible focal lengths for the lens and give the corresponding ray diagrams.

34 Two convex lenses, A and B, of focal lengths 10 and 15 cm respectively, are arranged coaxially at a distance of 20 cm apart. An object placed in front of A is viewed through the lens system. Find the nature and position of the image formed when the object is (*a*) at an infinite distance, (*b*) 20 cm, (*c*) 10 cm, (*d*) 5 cm in front of A.

35 What do you understand by the 'equivalent lens' of a system of thin lenses? Derive an expression for the focal length of the equivalent lens of two thin lenses of focal lengths f_1 and f_2 situated a distance a apart.

Two convex lenses, each of focal length f, are placed a distance of

$4f$ apart. Find the range of object distances for which a real image is formed by the combination.

36 Prove the formula $xx' = f^2$ for a thin lens where x and x' are the distances of the object and image from their corresponding focal points, and f is the focal length of the lens.

An illuminated linear object, 5 cm long, is positioned in front of an inaccessible lens, a real image of the object being received on a screen placed behind the lens. A movement of the object necessitates a displacement of the screen 10 cm further away from the lens, and it is observed that the image size has increased by 2 cm. Calculate the focal length of the lens and the original position of the object.

37 A convex lens of focal length 15 cm is placed on a plane mirror at the bottom of a gas-jar into which water is now poured to a depth of 10 cm. Find the position of an object, placed above the water surface, in which it is in coincidence with its image. Assume the refractive indices of water and of the lens material to be 4/3 and 3/2 respectively.

38 Establish the expression for the focal length of a thin lens in terms of the radii of its surfaces and the refractive index of the lens material.

A watch glass containing a little water is placed on a horizontal mirror and a pin held vertically above the watch glass coincides with its image when at a distance of 24 cm from the liquid lens. Taking the refractive index of water as 4/3, find the radius of curvature of the surface of the watch-glass.

39 A small illuminated aperture is moved towards a biconvex lens behind which is placed a plane mirror. Images of the aperture are formed alongside it when it is 20·0 cm, and again when it is 7·5 cm from the lens. On reversing the lens in its stand two similar images are formed, the corresponding distances being 20·0 cm and 15·0 cm. Explain the formation of the images and calculate the refractive index of the lens.

Worked example

40 *Light from a small, brightly-illuminated aperture is directed on to an equi-biconvex lens and an image is thrown back on the surface of the screen surrounding the aperture when the latter is at a distance of* 10 *cm from the lens. Explain the formation of this image and calculate the focal* *length and refractive index of the lens given that the radius of curvature of the lens surfaces is 20 cm.*

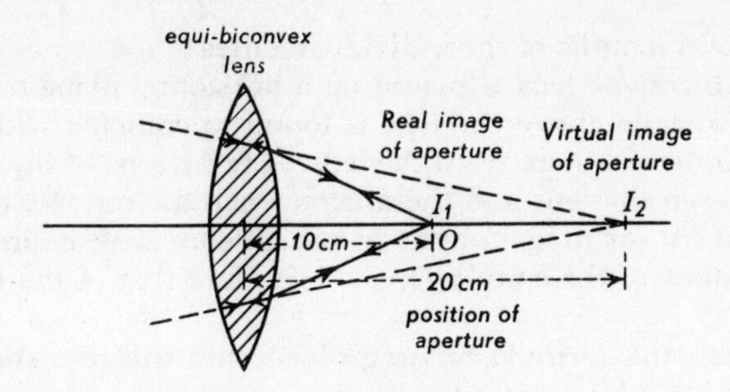

Fig. 26

The image on the screen surrounding the aperture is formed by rays from the aperture which, after being refracted at the first surface of the lens, strike the back surface normally and the partially reflected light thus retraces its path to give an image in the plane of the aperture (see diagram). The bulk of the light, of course, proceeds through the lens without further change of direction, proceeding as if emerging from the point I_2. Thus O and I_2 are respectively (real) object and (virtual) image points for refraction through the lens as a whole. Hence, using the lens formula $\frac{1}{u}+\frac{1}{v} = \frac{1}{f}$, and remembering that I_2 is positioned at the centre of curvature of the back surface (i.e. 20 cm from the lens), we have

$$\frac{1}{10} + \frac{1}{-20} = \frac{1}{f}$$

from which $f = \underline{20}$ cm

Now the focal length of the lens is related to the refractive index (n) of its material and the radii (r_1, r_2) of its surfaces by the formula

$$\frac{1}{f} = (n-1)\left(\frac{1}{r_1}+\frac{1}{r_2}\right)$$

In this case $r_1 = r_2 = 20$ cm (both being positive surfaces), and hence

$$\frac{1}{20} = (n-1)\frac{2}{20}$$

giving $n = \underline{1{\cdot}5}$

41 Prove that the focal length (F) of a combination of two thin lenses in contact can be obtained from the relation $\frac{1}{F} = \frac{1}{f_1} + \frac{1}{f_2}$ where f_1 and f_2 are the focal lengths of the individual lenses.

An equi-biconvex lens is placed on a horizontal plane mirror and a pin held vertically above the lens is found to coincide with its image when positioned 20·0 cm from the lens. A few drops of liquid are now placed between the lens and the mirror when the pin has to be raised a further 10 cm for it again to be in coincidence with its image. If the refractive index of the convex lens is 1·50, find that of the liquid.

42 Establish the formula relating object and image distances when refraction takes place at a single spherical surface.

The greatest thickness of a plane-convex lens is 3·0 mm. When viewed through its plane surface this thickness appears to be 1·95 mm, and when viewed through its curved surface, 2·06 mm. Use this data to calculate (i) the refractive index of the glass of the lens, (ii) the radius of curvature of the curved surface.

43 An illuminated point object is placed 10 cm from the centre of a glass sphere of radius 5 cm. If the refractive index of the glass is 1·5, find the nature and position of the image produced by the sphere.

44 A glass sphere of refractive index 1·5 has a small air bubble in it 2 cm from the centre. What is the apparent position of this bubble when viewed (i) through the side of the sphere nearest the air bubble, (ii) through the opposite side? (It is assumed that the viewing axis passes through the air bubble and the centre of the sphere).

45 The plane surface of a thin plano-convex lens is silvered and a point object placed on the axis of the system in front of the curved surface. Deduce a relationship between the distance of the object and that of the image formed by the system. If the refractive index of the lens is 1·5 and the radius of curvature of the curved surface is 20 cm, calculate the focal length of the 'equivalent concave mirror' so formed.

46 Give the details of Boys' method of determining the radius of curvature of a lens surface.

A thin plano-convex lens is floated on mercury with its plane surface in contact with the mercury, and it is found that an object coincides with its reflected image when it is 30 cm above the lens. The lens is now reversed and in order again to obtain coincidence of object and image, the object has to be moved 20 cm nearer the lens. Find the focal length of the lens and its refractive index.

47 Light from an illuminated point object on the axis of a biconvex lens undergoes two internal reflections before emerging from the lens. Deduce a relationship between the distance from the lens of the image so formed, the object distance, the refractive index of the lens and its principal focal length. Hence show that the 'secondary' focal length of a lens for which the refractive index is 1·5 is one-seventh of the principal focal length.

48 A hemisphere of transparent plastic material is used with its flat surface down as a magnifying-glass for map-reading. Assuming the eye to be positioned vertically above the curved surface of the lens on its axis of symmetry and at a distance d from the map, show that for the entire surface of the map covered to be visible the limiting value of d is given by the expression $\dfrac{rn^2}{2\sqrt{n^2-1}}$.

Hence, or otherwise, determine the least value of n for the lens material for which the full map area covered by the lens may be viewed through it.

49 When viewed from a distance the mercury thread of a mercury-in-glass thermometer appears to be half as thick as the stem. Taking the refractive index of the glass to be 1·5, calculate the actual diameter width of the mercury thread if the external diameter of the stem is found, on measurement, to be 3·0 mm.

50 An object is situated 6 m from a converging lens of focal length 10 cm. Behind the converging lens, and 6 cm from it, is placed a diverging lens of focal length 12 cm. What is the nature, position, and relative size of the final image produced by the lens combination?

51 If, in the above problem, the convex lens were used alone, what would be its position when giving an image (*a*) in the same position, (*b*) of the same size, as that produced by the lens combination of the given object.

52 Describe methods for finding the focal length of a concave lens using a suitable convex lens (*a*) in contact, (*b*) not in contact with it.

An eye placed 10 cm behind a concave lens of focal length 30 cm views a wall 120 cm from the lens. If the diameter of the lens is 5 cm, what length of wall can be seen through the lens?

Dispersion by prisms and lenses

In questions 53 to 56 (incl.) use the following refractive index values:

	Red (C)	Yellow (D)	Blue (F) light
Crown glass	1·515	1·517	1·523
Flint glass	1·644	1·650	1·664

53 Define *dispersive power*.

Calculate the dispersive power of a crown glass prism and the angular width of the spectrum produced by it if its refracting angle is 10°.

54 Calculate the angle of the flint glass prism which, combined with a 10° crown glass prism, produces a non-deviating combination. Find also the angular dispersion given by the prism combination.

55 Show that it is possible to produce an achromatic prism combination from separate crown and flint glass prisms.

What must be the angle of a flint glass prism which produces achromatic condition for the C and F range in the hydrogen spectrum with a crown glass prism of refracting angle 10°? What is the angular deviation obtained with the achromatic prism?

56 Calculate the separation of the foci for red and blue light when using a convex lens of mean focal length 1 m. How is the result related to the dispersive power?

Worked example

57 *Show how to make an achromatic doublet of focal length* 200 *cm using the glasses listed below:*

Crown glass		*Flint glass*	
n_{red}	n_{blue}	n_{red}	n_{blue}
1·515	1·523	1·644	1·664

Calculate the focal lengths of the two components.

The achromatic doublet consists of a convex lens (focal length f_1) of crown glass (dispersive power ω_1) in contact with a concave lens of focal length f_2 of flint glass (dispersive power ω_2). If the focal length of the combination is F, then, for two thin lenses in contact,

$$\frac{1}{f_2}+\frac{1}{f_2}=\frac{1}{F}=\frac{1}{200} \tag{1}$$

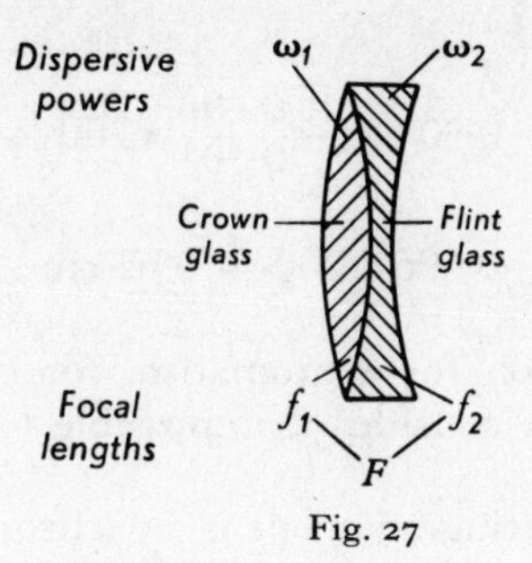

Fig. 27

The condition for achromatism is

$$\frac{\omega_1}{f_1} + \frac{\omega_2}{f_2} = 0$$

or

$$\frac{f_1}{f_2} = -\frac{\omega_1}{\omega_2}$$

Now

$$\omega_1 = \frac{1{\cdot}523 - 1{\cdot}515}{1{\cdot}519 - 1} \quad \text{and} \quad \omega_2 = \frac{1{\cdot}664 - 1{\cdot}644}{1{\cdot}654 - 1}$$

$$= \frac{0{\cdot}008}{0{\cdot}519} \qquad\qquad = \frac{0{\cdot}020}{0{\cdot}654}.$$

(taking the refractive index for yellow light as being the arithmetical mean of the refractive indices for red and blue light in each case). Thus

$$\frac{f_2}{f_2} = -\frac{\frac{0{\cdot}008}{0{\cdot}519}}{\frac{0{\cdot}020}{0{\cdot}654}} = -0{\cdot}504$$

or

$$f_1 = -0{\cdot}504 f_2 \tag{2}$$

Inserting this value of f_1 in equation (1), we have

$$\frac{1}{-0{\cdot}504 f_2} + \frac{1}{f_2} = \frac{1}{200}$$

i.e.

$$-\frac{0{\cdot}496}{0{\cdot}504 f_2} = \frac{1}{200}$$

or

$$f_2 = -\frac{0{\cdot}496}{0{\cdot}504} \times 200 = -\underline{196{\cdot}8} \text{ cm}$$

and

$$f_1 = -0{\cdot}504 \times -\frac{0{\cdot}496}{0{\cdot}504} \times 200$$

$$= 0{\cdot}496 \times 200 = \underline{99{\cdot}2} \text{ cm}$$

58 Establish the condition for achromatism for two thin lenses in contact. Why did Newton consider it impossible to construct achromatic lenses?

The objective of a refracting telescope is an achromatic combination of a crown glass lens and a flint glass lens. If the focal length of the combination is 1 metre, find the focal lengths of the component lenses. Use the following table of refractive indices.

	Crown glass	*Flint glass*
Red light	1·51	1·64
Blue light	1·53	1·68

59 Show how it is possible to produce a combination of prisms which produce dispersion without deviation.

Give a diagram, and a brief description, of a direct vision spectroscope.

60 Describe, with a suitable diagram, the optical system of the prism spectrometer. What adjustments are needed before measurements are taken with such an instrument?

Rainbows

61 Give an account of the formation of rainbows. Under what conditions would a completely circular bow be seen? Describe briefly some other optical phenomena in meteorological physics.

62 Assuming the refractive index of water for red light to be 1·33, calculate the angle subtended at the eye of an observer by the red arc of a primary bow.

63 Using the data given below, find the minimum deviations for red and violet light undergoing two internal reflections in a spherical water drop. Hence find the angular width of a secondary bow. ($n_{\text{red}} = 1{\cdot}33$, $n_{\text{violet}} = 1{\cdot}34$).

The eye. Defects of vision

64 Draw a sectional diagram of the human eye labelling the parts of importance in its optical behaviour. Discuss the defects of vision giving a brief account of methods of correcting them.

65 Compare and contrast the human eye and the photographic camera from the point of view of the optical arrangements.

A short-sighted person cannot see distinctly objects placed at a distance greater than 5 metres from his eyes. What is the nature and focal length of the spectacle lenses he requires for viewing distant objects?

Worked example

66 *The range of vision of a short-sighted person extends between the distances of* 10 *cm and* 30 *cm from the eye. Calculate the focal length of the spectacle lenses which he must wear to see distant objects at his far point. With these spectacles, what would be the least distance of his distinct vision?*

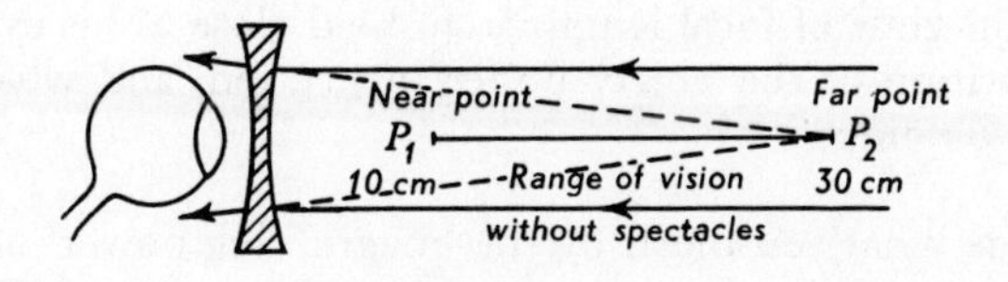

Fig. 28

For distant vision concave spectacles will be required. Parallel rays from the distant object must appear to diverge from the person's far point (30 cm) after refraction through the lens. Hence the focal length of these concave spectacle lenses must be 30 cm.

The nearest position for an object for clear vision with these spectacles is such that rays diverging from this point (distance u from the spectacles) appear to do so from P_1—the near point of the unaided eye. Hence, using the lens formula $\frac{1}{u}+\frac{1}{v} = \frac{1}{f}$, and inserting -10 for v (virtual image) and -30 for f (concave lens) we have

$$\frac{1}{u}+\frac{1}{-10} = \frac{1}{-30}$$

from which u, the least distance of distinct vision with the spectacles, is 15 cm.

67 Define the unit of power of a lens.

A person whose range of distinct vision is from 12·5 to 25·0 cm is provided with spectacles which remove his near point to 25·0 cm. What is the nature and power of his spectacle lenses and what is the greatest distance to which he can see distinctly when using the spectacles?

68 What is meant by *accommodated vision,* and to what is it due?

The range of distinct vision of an elderly person is from 75 to 200 cm. What spectacles will he require for (*a*) reading, (*b*) walking purposes, and what is his range of vision when wearing these spectacles? You may assume the normal reading position to be 25 cm from the eye.

Optical instruments

69 Describe, with the use of a suitable ray diagram, the use of a convex lens as a simple magnifying-glass. Derive a formula for its magnifying power.

A person whose nearest point of distinct vision is 30 cm uses a magnifying-glass of focal length 5 cm held close to his eye. What must be the position of the object under inspection, and what is the magnification obtained?

70 Define what you mean by the 'magnifying power' of a compound microscope and show that, for a given instrument, the magnifying power is approximately inversely proportional to the focal length of the objective lens.

Describe how you would determine the magnifying power of a compound microscope experimentally.

71 A compound microscope consists of an object glass of focal length 2·5 cm situated 30 cm in front of an eyepiece of focal length 7·5 cm. The object distance is adjusted so that an observer using the microscope sees the final image at a distance of 25 cm from the eye-lens. Compare the angle subtended by this image at the eye-lens with that subtended by the object when positioned 25 cm from the eye.

72 A compound microscope consists of an objective lens of focal length 1 cm and an eye lens of focal length 5 cm mounted in a tube at a distance of 20 cm from each other. Find where the object must be placed when viewed through the instrument by a person whose nearest point of distinct vision is 25 cm. What adjustment to the microscope would be required if it is then used by a person whose nearest point of distinct vision is 20 cm?

Worked example

73 *Thin converging lenses of focal lengths 2 cm and 9 cm are used respectively as the objective and eyepiece of a microscope, the centres of the lenses being 25 cm apart. If an object is placed at a distance of 2·25 cm from the objective, what will be the position and magnification of the final image?*

Objective lens. O is the object, I, the real image produced by it. Using the symbols shown in the diagram, we have,

$$\frac{1}{u_1}+\frac{1}{v_1}=\frac{1}{f_1} \quad \text{or, inserting numbers,} \quad \frac{1}{2\cdot 25}+\frac{1}{v_1}=\frac{1}{2}$$

from which $\dfrac{1}{v_1}=\dfrac{1}{2}-\dfrac{4}{9}$ giving $v_1 = \underline{18}$ cm

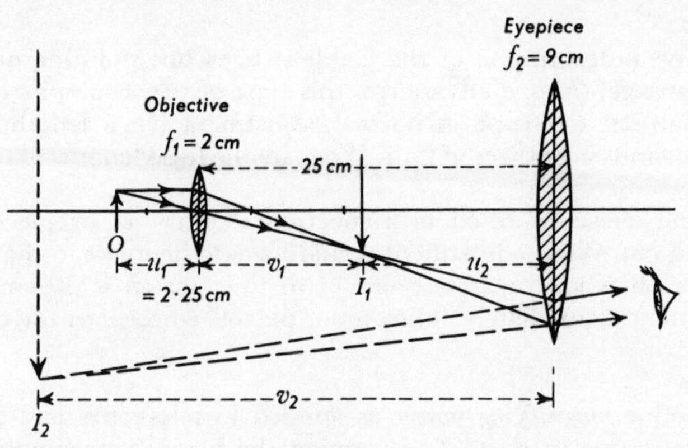

Fig. 29

Eyepiece. I_1 acts as object for the eyepiece which produces a virtual image at I_2. Now the object distance u_2 is clearly $25-18 = 7$ cm. Hence, since

$$\frac{1}{u_2}+\frac{1}{v_2}=\frac{1}{f_2}$$

we have

$$\frac{1}{7}+\frac{1}{v_2}=\frac{1}{9}$$

from which $\dfrac{1}{v_2}=\dfrac{1}{9}-\dfrac{1}{7}$ giving $v_2 = \underline{-31\cdot 5}$ cm

Thus the final image is a virtual image (indicated by the negative sign) 31·5 cm in front of the eye-lens.

Magnification. The linear size of I_1 is $\frac{v_1}{u_1}$ times that of the object, i.e. $I_1 = \frac{18}{2{\cdot}25} \times O = 8 \times O$ (where O represents the linear dimensions of the object).

The linear size of I_2 is $\frac{v_2}{u_2}$ times that of I_1,

i.e. $I_2 = \frac{31{\cdot}5}{7} \times I_1 = 4{\cdot}5 \times 8 \times O = \underline{36 \times O}$

Hence the magnification of the final image

$$= \frac{I_2}{O} = \underline{36}.$$

74 Give a description of the Galilean telescope pointing out (*a*) one disadvantage, (*b*) one advantage, this type of telescope possesses.

A Galilean telescope in normal adjustment has a length of 15 cm and a magnifying power of four. What are the focal lengths of the lenses?

75 The telescope fitted to a spectrometer has an eyepiece of focal length 5 cm. What adjustment would have to be made to the telescope if a person whose nearest point of distinct vision is 20 cm uses the instrument immediately following a person whose nearest point is at 25 cm?

76 Define *magnifying power* as applied to telescopes and show that for a telescope in normal adjustment the magnifying power is given by the ratio of the focal length of the objective lens to that of the eye-piece.

Describe a simple method of measuring the magnifying power of a small telescope in the laboratory.

77 Describe, with diagrams, reflecting telescopes of (*a*) the Newtonian, (*b*) the Cassegranian type. Why are large modern astronomical telescopes built as reflectors?

78 An astronomical telescope with an objective of focal length 100 cm and an eyepiece of focal length 3 cm is adjusted so as to throw a real image of the sun on a screen 20 cm from the eye-lens. If the diameter of this image is 4·9 cm, calculate the angle subtended by the sun at the centre of the objective of the telescope.

79 Give an account, with diagrams, of the optical system of a photographic camera. What are the essential requirements of a good photographic lens?

80 What do you understand by the description f/8 for the aperture of a photographic lens? Compare the exposure times needed for a given camera working at (i) f/8, (ii) f/11 on an object under the same conditions of illumination. What difference would be observable in the two pictures?

81 Describe the optical system of a projection lantern.

It is required to project pictures 1·5 m square on a screen situated 5 m from a projection lantern. If the slides are 3 cm square, what must be the focal length of the projection lens?

82 What is meant by *episcopic projection*? In what particulars must the ordinary projection lantern be modified for this purpose?

83 What do you understand by the term *resolving power* as applied to a telescope? Describe a simple laboratory method of finding the resolving power of a small telescope.

84 An astronomical telescope can just resolve two stars which have an angular separation of one second of arc. If the eye has a resolving power of two minutes of arc, what must be the magnifying power of the telescope if full advantage is to be taken of its resolving power?

If the aperture of the telescope were to be stopped down to half its original diameter, what would be the effect on (i) the resolving power, (ii) the magnifying power of the telescope and (iii) the brightness of the image formed?

Velocity of Light

85 Describe fully a method of measuring the velocity of light in air giving a diagram of the apparatus and indicating how the required velocity is calculated from the observations made.

Given that the velocity of light in air is 3×10^8 m s^{-1}, calculate its velocity in glass for which the refractive index is 1·50.

86 Calculate, for an observer on the earth, the time lag between the calculated and observed eclipse times of the moons of Jupiter for the period when the earth and Jupiter are in conjunction until when they are next in opposition. (Mean distance from the earth to the sun $= 149{\cdot}5 \times 10^6$ km; velocity of light $= 3 \times 10^8$ m s^{-1}.)

87 Give an account of Römer's method of finding the velocity of light, explaining how the value is calculated from the observations made.

88 If the mean interval between successive eclipses of Jupiter's second satellite is 42 hr 29 min, calculate the greatest and least intervals between the eclipses given that the mean distance from the sun to the earth is $149{\cdot}5 \times 10^6$ km, and that the velocity of light is 3×10^8 m s^{-1}.

89 Draw a diagram of Fizean's toothed-wheel apparatus for measuring the velocity of light. What are the chief difficulties met with in the experiment?

In such an experiment the distance between the wheel and the mirror was 20 km and the wheel had 100 teeth and 100 spaces of equal angular width. At what speed was the wheel rotating when the first eclipse of the image occurred? (Take the velocity of light as 3×10^8 m s^{-1}.)

90 Calculate the speed of the wheel in the above problem at which (*a*) the first reappearance, and (*b*) the third eclipse of the image occurs.

91 Describe Foucault's rotating mirror method for measuring the velocity of light. What are the main criticisms of this method, and how has the method been modified by later workers to meet these criticisms?

92 In a simple Foucault's apparatus for measuring the velocity of light the distance between the rotating plane mirror and the concave reflector was 1 km, whilst the distance between the rotating mirror and the observing eyepiece was 5 metres. Taking the velocity of light as 3×10^8 m s^{-1}, what was the speed of rotation of the plane mirror when the image had a displacement of 20 mm?

Photometry

93 What are the main difficulties encountered in simple visual photometry? Describe the details of design and the mode of operation of a modern photometer in which these difficulties have been overcome.

94 Define *lumen, luminous intensity* and *illumination of a surface.*

Two lamps with luminous intensities of 5 and 20 candela respectively, are placed 1 m apart and a screen is placed between them 40 cm from the weaker lamp. Compare the illumination at either side of the screen. Where must the screen be placed to be equally illuminated by both lamps?

95 Give a detailed account of how you would use a photometer to compare the luminous intensities of two lamps, of luminous intensities of 20 cd and 30 cd respectively, are situated 80 cm apart and an opaque white screen is placed mid-way between them. When must a 40 cd lamp be placed so that the screen is equally illuminated by both sides?

Worked example

96 *A lamp A,* 30 *cm from a photometer head, is in photometric balance with another lamp B on the other side of the head. How will the position of A have to be adjusted if a sheet of frosted glass, with a transmission coefficient of* 75 *per cent, is placed between B and the photometric head?*

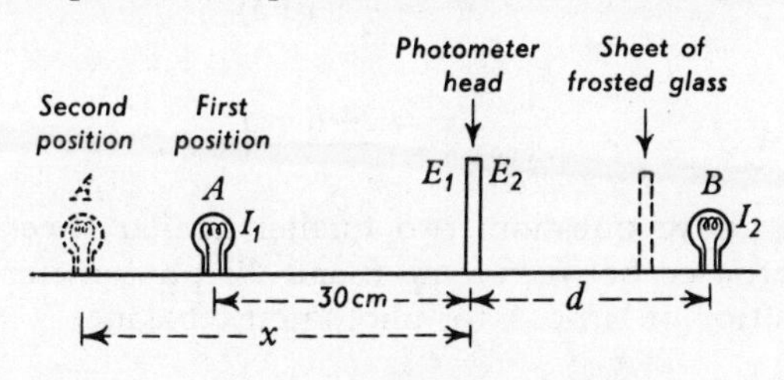

Fig. 30

Let I_1, I_2 be the luminous intensities of the lamps A and B respectively, and E_1, E_2 the illuminations they produce at the photometer head. Then, by the photometric relationship,

$$E = \frac{I}{(\text{distance})^2}$$

we have

$$E_1 = \frac{I_1}{(0{\cdot}3)^2} \quad \text{and} \quad E_2 = \frac{I_2}{d_2}$$

(d being the unknown distance of B from the photometer). But since the lamps are in photometric balance, $E_1 = E_2$, i.e.

$$\frac{I_1}{(0{\cdot}3)^2} = \frac{I_2}{d^2} \tag{1}$$

Interposing the frosted glass between the photometer and lamp B cuts off some of the luminous flux emitted by the lamp which now has an effective luminous intensity of $\frac{75}{100} I_2$ as regards the illumination at the photometer. Let lamp A be moved back to a position x cm from the photometer to re-establish photometric balance, then

$$\frac{I_1}{x^2} = \frac{75}{100} \frac{I_2}{d^2} \tag{2}$$

or

$$\frac{4}{3} \frac{I_1}{x^2} = \frac{I_2}{d^2} \tag{2a}$$

Hence from equations (1) and (2a) we have

$$\frac{I_1}{(0{\cdot}3)^2} = \frac{4}{3} \frac{I_1}{x^2}$$

giving

$$x^2 = \frac{4}{3} \times (0{\cdot}3)^2$$

from which

$$x = \underline{34{\cdot}6} \text{ cm}$$

97 If, in the above question, two further similar sheets of frosted glass were interposed between lamp B and the photometer head, what is now the position of lamp A for photometric balance?

98 Two lamps, A and B, are in photometric balance when placed either side of a photometric head and with lamp A 50 cm from the head. A thin sheet of frosted glass is now interposed between lamp B and the photometric head when it is found that lamp A has to be moved 20 cm to restore balance. Calculate the transmission factor of the glass.

99 A lamp is positioned 5 m vertically above a work bench. Where must a plane mirror be placed to increase by 50 per cent the illumination of the bench immediately below the lamp? (You may assume the mirror to have a perfectly reflecting surface.)

100 Two lamps provide a photometric balance when positioned 25 cm and 75 cm respectively on either side of a screen. A polished surface is now placed 10 cm behind the weaker lamp, when it is found that the other lamp has to be moved 5 cm to restore the balance. Calculate the reflection factor of the polished surface.

101 Derive an expression showing how the illumination of a surface depends upon its inclination to the incident light.

A lamp is situated 2 m above the centre of a circular table of radius 1·5 m. Compare the illumination at the centre of the table with that at its edge.

102 A street is lit by lamps supported on standards 10 m high situated at 40 m intervals from each other. If the minimum pavement illumination along the line of the lamps is to be 0·5 lux, obtain an approximate value for the luminous intensity of the lamp heads.

Wave theory

103 What is meant by the term *interference* in optics? State the conditions necessary for the effects of interference to be observed, and give an account of some method, based on these effects, of measuring the wavelength of light.

104 One edge of a microscope slide 3 cm long is in contact with a plane glass plate, the other edge being raised from the plate by inserting a strip of paper under it. The air wedge so formed between the slide and the plate is now illuminated by sodium light incident normally from above, and in viewing the reflected light parallel interference bands are observed at regular spacings of 0·20 mm. Account for these fringes and calculate the thickness of the paper, assuming a value for the wavelength of sodium light of $5{\cdot}89 \times 10^{-7}$ m.

105 Describe the production of interference fringes by means of Fresnel's biprism.

In such an experiment the width of the fringes was found to be 0·320 mm when viewed through a microscope positioned with its focal plane 82·5 cm from the illuminated slit. On now interposing a convex lens between the biprism and the microscope, two positions of the lens were found in which double images of the slit were in focus in the focal plane of the microscope. In one position these images were 4·525 mm apart, and in the other position 0·510 mm apart. Find the wavelength of the light, assuming it to be monochromatic.

106 Two plane mirrors, inclined at an angle of 179° with each other, are symmetrically placed with their reflecting surfaces towards a narrow slit which is 10 cm distant from the common edge of the mirrors. The slit is now illuminated with sodium light of wavelength $5{\cdot}893 \times 10^{-7}$ m and interference fringes are observed on a screen 1 m from the slit. Draw a diagram of the arrangement and calculate the width of the fringes.

107 When a thin plane parallel piece of glass of refractive index 1·52 was interposed perpendicularly in the path of one of the interfering pencils of a Young's double slit arrangement, the centre of the fringe system was displaced through 20 fringe widths. If the wavelength of the light used was $5{\cdot}893 \times 10^{-7}$ m, what was the thickness of the glass?

108 How can the wavelength of sodium light be determined from observations made on Newton's rings? Sketch the arrangement of the apparatus and derive the working formula you would use.

Monochromatic light falls normally on a plano-convex lens placed with its curved surface in contact with an optically plane glass plate. Viewed by reflected light the diameter of the 10th dark ring of the resulting system of interference fringes is found to be 0·485 cm. If the radius of the curved surface of the lens is 1 m, what is the wavelength of the light used in the experiment?

109 The space between a plano-convex lens and the plane glass plate on which it is placed is filled with a liquid. On viewing the system of interference fringes formed by the reflection of sodium light incident normally on the lens from above it is found that the diameter of the 20th dark fringe is 0·84 cm. If the radius of the curved face of the lens is 2 m, and the wavelength of sodium light is $5{\cdot}89 \times 10^{-7}$ m, find a value for the refractive index of the liquid.

110 A thin coating of Canada balsam on the surface of a sheet of crown glass is illuminated with white light at 60°. On examining the reflected light spectroscopically a dark band is observed in the region corresponding to a wavelength of 6×10^{-7} m. Explain this and calculate the thickness of the film given that the refractive index of Canada balsam is 1·53.

111 Describe the use of a diffraction grating for measuring wavelengths.

A parallel beam of sodium light of wavelength $5{\cdot}89 \times 10^{-7}$ m is incident normally on a grating, and the first order diffracted image is formed at an angle of $26°13'$ with the normal. Calculate the number of lines per centimetre ruled on the grating.

112 A diffracting grating having 5000 rulings per centimetre is illuminated normally by a parallel beam of sodium light. Calculate (*a*) the highest order spectrum that can be seen, and (*b*) the angular separation of the D_1 and D_2 lines in the second order spectrum if the wavelengths of these lines are respectively $5{\cdot}896 \times 10^{-7}$ and $5{\cdot}890 \times 10^{-7}$ m.

Worked example

113 *A diffraction grating with* $6{\cdot}3 \times 10^5$ *lines per metre is set on the table of a spectrometer so that monochromatic light is incident normally on it. The telescope is moved round and bright images appear on the cross-wires at the following settings:* 80° 54′; 107° 28′; 130° 00′; 152° 12′; 179° 6′. *Explain what is being observed at each of these settings, and calculate the wavelength of the light.*

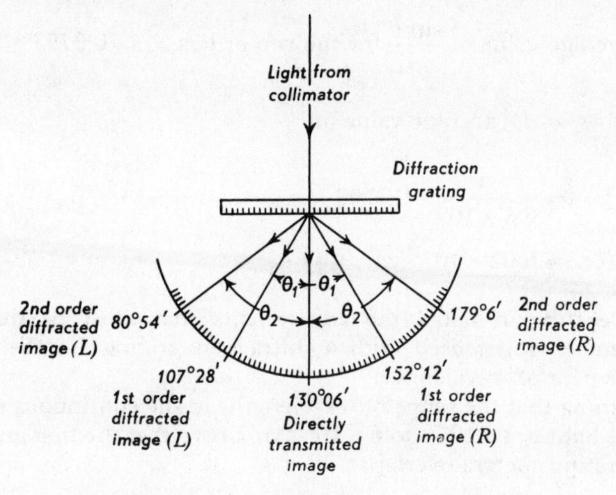

Fig. 31

Diffracted images of the collimator slit are observed at the various settings of the telescope as illustrated in the diagram. Now by the theory of the diffraction grating,

$$\sin \theta = \frac{n\lambda}{d}$$

Where θ is the angle on either side of the normal for the formation of the nth order diffracted image for light of wavelength λ when the grating element is

$$d = \frac{1}{(\text{no. of lines per m})}$$

Hence $$\lambda = d\,\frac{\sin\theta}{n}$$

Now, from the observations, the mean angle of diffraction for the first order ($n = 1$) is $\dfrac{152°\,12' - 107°\,28'}{2} = 22°\,22'$ and the mean angle of diffraction for the second order ($n = 2$) is $\left(\dfrac{179°\,8' - 80°\,54'}{2}\right) = 49°\,6'$.

Hence $\dfrac{\sin\theta}{n}$ for first order image $= \dfrac{\sin 22°\,22'}{1} = 0{\cdot}3806$

and $\dfrac{\sin\theta}{n}$ for second order image $= \dfrac{\sin 49°\,6'}{2} = 0{\cdot}3780$

Thus average value of $\dfrac{\sin\theta}{n}$ for the two orders $= 0{\cdot}3793$

$$\begin{aligned}\text{Now}\quad \lambda &= d \times \text{average value of } \frac{\sin\theta}{n}\\ &= \frac{1}{6{\cdot}3\times10^{5}}\times 0{\cdot}3793\\ &= \underline{6{\cdot}02\times10^{-7}\ \text{m}}\end{aligned}$$

114 Describe in detail the adjustments that must be made to a spectrometer when used with a diffraction grating for the accurate measurement of wavelength.

Assuming that the range of wavelengths in the continuous spectrum of white light is 4×10^{-7} to 8×10^{-7} m, prove that the first and second order grating spectra overlap.

115 Discuss the phenomenon of diffraction at a straight edge.

A uniform vertical wire 0·5 mm in diameter is positioned 15 cm in front of a narrow vertical slit illuminated with monochromatic light. A shadow of the wire is thrown on a screen 1 metre from the wire and a number of light and dark bands of width 0·118 cm are observed inside the shadow. Explain the origin of these bands and calculate a value for the wavelength of the light used. How many bright bands will be seen in the shadow zone under the conditions of the experiment?

116 The separation of the crystal planes in a rock salt crystal is $3{\cdot}0\times10^{-10}$ m. When this crystal is used in an X-ray spectrometer the glancing angle for the first order spectrum was found to be 16·5°. Calculate the wavelength of the X-rays used.

117 What do you understand by the polarization of light. What information does it provide concerning the wave nature of light?

118 Describe two ways of producing plane polarized light and discuss briefly two uses of polarized light.

119 What do you understand by *circularly polarized light*? How is it produced, and how can it be distinguished from unpolarized light?

120 What is a *quarter wave plate* and for what purpose is it used?

Given that the ordinary and extraordinary refractive indices of quartz are 1·544 and 1·553 respectively, calculate the minimum thickness for a quarter wave plate of quartz when used with sodium light of wavelength $5{\cdot}89 \times 10^{-7}$ m.

121 Discuss the production of polarized light by reflection and define the term *polarizing angle* used in this reference. Calculate the value of this angle for crown glass of refractive index 1·52. Give the details of some practical device for producing strong polarized light beams by reflection.

SOUND

Sound waves. Characteristics of musical sound

1 Describe experiments to show that sound waves may be (i) reflected, (ii) refracted. Discuss examples of the reflection and refraction of sound in nature.

2 What do you understand by (*a*) *progressive waves,* (*b*) *nodes* and *antinodes*? Discuss the interference of sound waves with particular reference to the formation of (i) beats, (ii) standing vibrations.

3 State what you understand by the term *wave,* and obtain a general equation representing progressive wave motion in a medium.

The equation below represents (in SI units) the motion of a sound wave travelling through a gas:

$$y = 10^{-6} \sin 2\pi\,(1{\cdot}47x - 500\,t).$$

What information concerning the sound wave can you deduce from this equation?

4 Show that if two notes of frequencies f_1 and f_2 are sounded together, the frequency of the resulting 'beats' is equal to the difference between f_1 and f_2.

Four beats per second can be heard when a tuning-fork of frequency 256 Hz, and a wire of length 1 m, vibrating transversely, are sounded together. The vibrating length of the wire is now reduced to 90 cm and 8 beats per second are heard when the wire is sounded together with a tuning-fork of frequency 288 Hz. What was the original frequency of the note emitted by the wire? (Assume the wire to be in constant tension throughout.)

5 What are the frequencies of the notes on the diatonic scale if the frequency of the tonic (lowest note) is 128 Hz? Mention any difficulties experienced when using this scale with a keyed instrument, e.g. a piano. How is the scale modified to meet these difficulties?

6 In what way do musical tones differ from one another? Give a physical account of these differences.

7 Describe the 'falling plate' method of determining the frequency of a tuning-fork explaining how the result is calculated from the observations.

In such an experiment successive trains of 20 waves occupied lengths of 20·3 mm and 35·5 mm. Assuming the acceleration of gravity to be 9·8 m s^{-2}, find the frequency of the vibrating fork. The fork was then taken from its clamp and sounded, with stylus attached, simultaneously with a similar, but unloaded, fork when 4 beats per second were heard. What is the true frequency of the fork?

8 Describe the method of determining the frequency of a tuning-fork using a stroboscopic disc.

A disc marked with 100 evenly spaced radial slots and revolving at 300 revolutions per minute appears stationary when viewed through two overlapping slits attached to the prongs of a vibrating tuning-fork. What is the frequency of the fork? If 30 beats are counted in 5 seconds when the fork is sounded together with an identical fork (but without attached slits), what is the true frequency?

9 How are the loudness and the energy of a sound related? Discuss briefly the decibel scale of loudness.

The loudspeaker of a public address unit produces a noise level of 80 decibels at a point 5 m away. What will be the loudness in decibels at this point if the power delivered to the speaker is doubled?

Worked example

10 *The sound from a pneumatic drill gives a noise level of* 90 *decibels at a point a few metres from it. What is the noise level at this point when four such drills are working at the same distance away?*

On the decibel scale of loudness, the increase of loudness when the energy associated with the sound increases from E_1 to E_2

$$= 10 \log_{10} \frac{E_2}{E_1} \text{ decibels (db)}$$

Hence if the energy associated with one pneumatic drill is E, then, with four drills working, the increase in noise level

$$= 10 \log_{10} \frac{4E}{E}$$

$$= 6{\cdot}02 \text{ db}$$

Hence the new noise level at this point in question

$$= 90 + 6{\cdot}02$$

$$= \underline{96{\cdot}02} \text{ db}$$

11 An observer, at a given distance from the centre of an explosion, hears the report (*a*) by sound waves received directly, and (*b*) by sound waves reflected from a wall situated at an equal distance from the centre of the explosion but on the side remote from the observer. Compare the energies of the two sounds received by the observer, expressing your results in decibels. Assume the wall has a reflection coefficient of 50 per cent.

Velocity of sound

12 Give an account of the determination of the velocity of sound by 'free air' methods. What are the particular difficulties in these experiments, and how are they overcome?

13 How does the velocity of sound in air vary with (i) the pressure, (ii) the temperature, and (iii) with the humidity?

Given that the velocity of sound at s.t.p. is 330 metres per second, calculate its value on a day when the temperature is 20°C and the barometric height is 75 cm of mercury.

14 Account for the discrepancy between Newton's calculated value of the velocity of sound in air and the experimentally derived value.

Calculate the value of the velocity of sound in air using the following data: Density of air = 1·293 kg m^{-3}, barometric height on the particular occasion = 76 cm of mercury, relative density of mercury = 13·6, acceleration of gravity = 9·81 m s^{-2}, ratio of specific heat capacity at constant pressure to that at constant volume for air = 1·40.

15 Describe methods by which the velocity of sound in water has been determined.

Determine by how much the volume of 1 cm^3 of water changes with an increase of pressure of 1 atmosphere, using the fact that the velocity of sound in water is 1430 m s^{-1}.

Worked example

16 *The product of the pressure and volume of* 1 *kg of air at* 0°*C is* 7·84 × 10^4 *J, and the ratio of the specific heats is* 1·4. *Calculate the velocity of sound in air at* 15°*C.*

The expression for the velocity (c) of sound in a gas is

$$c = \sqrt{\frac{\gamma p}{\rho}}$$

where p is the pressure of the gas, ρ its density, and γ is the ratio of the specific heat capacities of the gas. Now density is $\frac{\text{mass}}{\text{volume}}$, hence if v m^3 is the volume of 1 kg of the gas, $\rho = \frac{1}{v}$, and accordingly the expression for the velocity becomes

$$c = \sqrt{\gamma p v}$$

Thus
$$c = \sqrt{1 \cdot 4 \times 7 \cdot 8 \times 10^4}$$
$$= \underline{3 \cdot 313 \times 10^2 \text{ m s}^{-1}}$$

This is the velocity at 0°C, however. The velocity at a temperature t°C is given by the relation

$$c_{t°\text{C}} = c_{0°\text{C}} \left(1 + \frac{t°\text{C}}{273}\right)^{\frac{1}{2}}$$

Hence
$$c_{15°\text{C}} = 3 \cdot 313 \times 10^2 \left(1 + \frac{15}{273}\right)^{\frac{1}{2}}$$
$$= 3 \cdot 313 \times 10^2 \left(\frac{288}{273}\right)^{\frac{1}{2}}$$
$$= \underline{3 \cdot 41 \times 10^2 \text{ m s}^{-1}}$$

17 A sound signal generated at the stern of a boat 80 metres long is received at the bow (i) by direct transmission through the air, and (ii) after reflection at the ocean bed. If the time interval between the two signals is 0·4 second, calculate the depth of the water below the boat, assuming the velocity of sound in air to be 333 m s^{-1} and in sea water 1450 m s^{-1}.

18 Describe an echo method for finding the velocity of sound in air.

An observer with a metronome finds that when at a distance of 105 m from a wall the echoes of the metronome's beat cease to be distinguishable. Accurate observations made subsequently with the metronome showed that the time interval between the 'ticks' was 0·63 s. What was the velocity of sound in air on the occasion?

19 A football on a path leading to a flight of stone steps with a depth of tread 0·3 m produces a regular note the pitch of which is identified as the second higher octave of a D tuning-fork (frequency 287). What result do these observations give for the velocity of sound in air?

20 A man approaching a distant wall (along a direction at right angles to the wall) makes a sharp sound and hears an echo 2·1 seconds later. He now continues walking towards the wall and, after traversing a further distance of 50 m, makes another sharp sound, the echo now being received 1·8 seconds later. Find the velocity of sound at the time and the man's original distance from the wall.

21 Using the data below, determine the velocity of sound in air on a day when the barometer reads 76 cm of mercury, the air temperature is 20°C, and the relative humidity is 100 per cent.

Velocity of sound in dry air at 0°C = 331 m s^{-1}.

Saturated vapour pressure of water at 20°C = 17·5 mm of mercury.

Density of water vapour is five-eighths that of dry air under the same conditions.

(You may assume that the value of γ for air is unchanged by the admission of water vapour.)

Vibrations in gas columns

22 Discuss, with suitable diagrams, the modes of vibration of closed and open pipes.

The second overtone of a closed organ pipe is found to be in unison with the first overtone of an open pipe. Ignoring end effects, compare the lengths of the two pipes.

Worked example

23 *In an experiment with a resonance tube the first two successive positions of resonance occurred when the lengths of the air columns were* 15·4 *cm and* 48·6 *cm respectively. If the velocity of sound in air at the time of the experiment was* 340 *m* s^{-1}, *calculate the frequency of the source employed and the value of the end correction for the resonance tube. If the air column is further increased in length, what will be the length when the next resonance occurs?*

The vibrations in the tube for the 1st and 2nd resonances are as shown in the diagrams. From these it will be clear that

$$\frac{\lambda}{4} = l_1 + e = 15{\cdot}4 + e \qquad (1)$$

(where e is the 'end correction)

and
$$\frac{3\lambda}{4} = l_2 + e = 48{\cdot}6 + e \qquad (2)$$

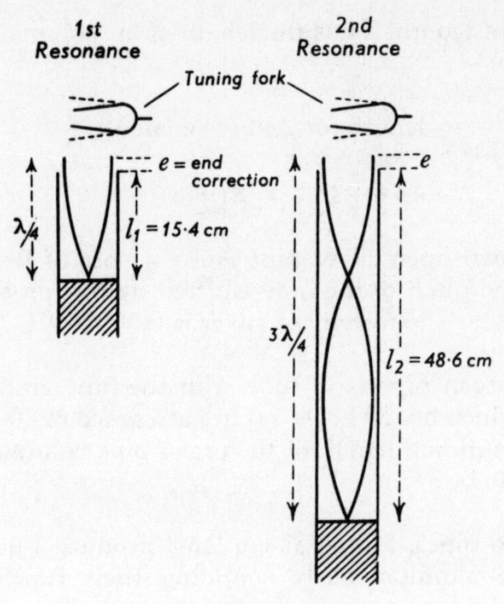

Fig. 32

Subtracting (1) from (2) we have

$$\frac{\lambda}{2} = l_2 - l_1 = 48{\cdot}6 - 15{\cdot}4 = 33{\cdot}2$$

giving $\lambda = 66{\cdot}4$ cm

Now, velocity = frequency × wavelength

Hence the frequency of the vibrating air column (which is also the frequency of the vibrating source)

$$= \frac{\text{velocity (m s}^{-1})}{\text{wavelength (m)}} = \frac{340}{0{\cdot}664}$$

$$= 512 \text{ Hz}$$

Knowing the value of λ, the 'end-correction' of the tube can be found from equation (1) above. Thus

$$\frac{66{\cdot}4}{4} = 15{\cdot}4 + e$$

from which $l = \underline{1{\cdot}2}$ cm.

The 3rd resonance occurs when the air column in the tube is long enough to accommodate half a wavelength more than is the case for the 2nd resonant length. Thus the length of air column in the tube for the 3rd resonance

$$= \text{length for 2nd resonance} + \frac{\lambda}{2}$$

$$= 48{\cdot}6 + 33{\cdot}2 = \underline{81{\cdot}8 \text{ cm}}$$

24 An air-blown open silver pipe emits a note of frequency 512 at 0°C. What is the pitch of the note emitted by this pipe at 20°C if the coefficient of linear expansivity of silver is $0{\cdot}000\,019\,°C^{-1}$?

25 A closed organ pipe is in tune with the note emitted by a siren when its disc, which has 30 holes, rotates at a speed of 800 revs. per min. Calculate the sounding length of the organ pipe assuming the velocity of sound in air to be 340 m s^{-1}.

26 Two closed pipes, 51 and 52 cm long, produce 3 beats per second when they are simultaneously sounding their fundamental notes. Ignoring end corrections, calculate a value for the velocity of sound in air from the above information.

Worked example

27 *A tube 40 cm long has a tight-fitting piston at one end and a cork at the other. The piston is gradually pushed in until, at a pressure of $1\frac{1}{2}$ atmospheres, the cork is expelled with a 'pop'. Calculate the frequency of the 'pop' given that the velocity of sound in air at the time* $= 340$ *m s*$^{-1}$.

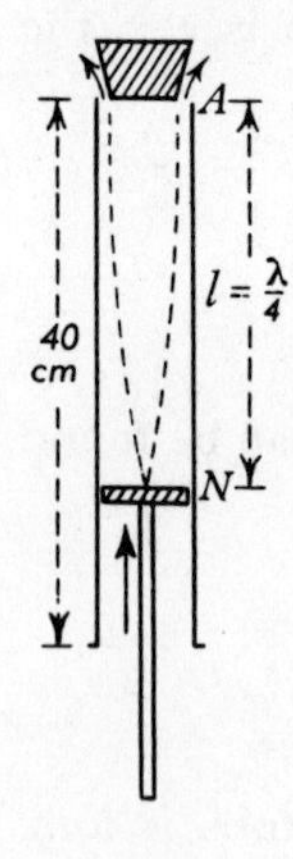

Fig. 33

Let l be the length of the air column in the tube when the cork 'pops'. The sudden release of the pressure sets the air in the tube in vibration with a node at the piston face and an antinode at the mouth of the tube where the air is in violent movement. Thus, ignoring end effects, the state of vibration of the air in the tube will be as indicated in the diagram and accordingly the wavelength (λ) of the vibration will equal $4l$. Now assuming the compression of the gas to take place isothermally from an initial pressure of 1 atmosphere, and taking the volume of the air in the tube as proportional to the length of it above the piston, we have, by Boyle's law

$$1 \times 40 = \frac{3}{2} \times l$$

from which $l = \frac{80}{3} = 26\frac{2}{3}$ cm

Thus $\lambda = 4l = \underline{106\frac{2}{3} \text{ cm}}$

Now if f is the frequency of the 'pop', we have,

$$\text{Velocity of sound in air} = f \times \lambda$$

Thus $340 \times 100 = f \times 106\frac{2}{3}$

or $f = \frac{340 \times 100}{106\frac{2}{3}} = \underline{318{\cdot}75 \text{ Hz}}$

28 An adjustable resonance tube, closed at one end, resounds to a fork of frequency 480 Hz held above the open end of the tube when the length of the air column in the tube is 16·9 cm and again when it is 52·1 cm. Calculate a value for the velocity of sound in air under the conditions of the experiment.

29 When a vibrating tuning-fork of frequency 256 Hz is held above the mouth of a closed resonance tube of adjustable length, the first position of resonance is obtained with an air column of length 31·3 cm. The corresponding length for a fork of frequency 320 Hz is 24·8 cm. Calculate from these observations the velocity of sound in air and the end correction of the tube.

Worked example

30 *Determine, neglecting end corrections, the lengths of the three shortest closed pipes, and of the three shortest open pipes, which would resound to a note of frequency* 512 *Hz when the air in the pipes is at a temperature of* 25°C. *(Velocity of sound in air at* 0°C = 330 *m* s^{-1}.*)*

Velocity of sound at 25°C

$$= \text{Velocity at } 0°\text{C} \left(1 + \frac{25}{273}\right)^{\frac{1}{2}}$$

$$= 330 \left(\frac{298}{273}\right)^{\frac{1}{2}}$$

$$= \underline{344{\cdot}7 \text{ m s}^{-1}}$$

Now Velocity = frequency × wavelength

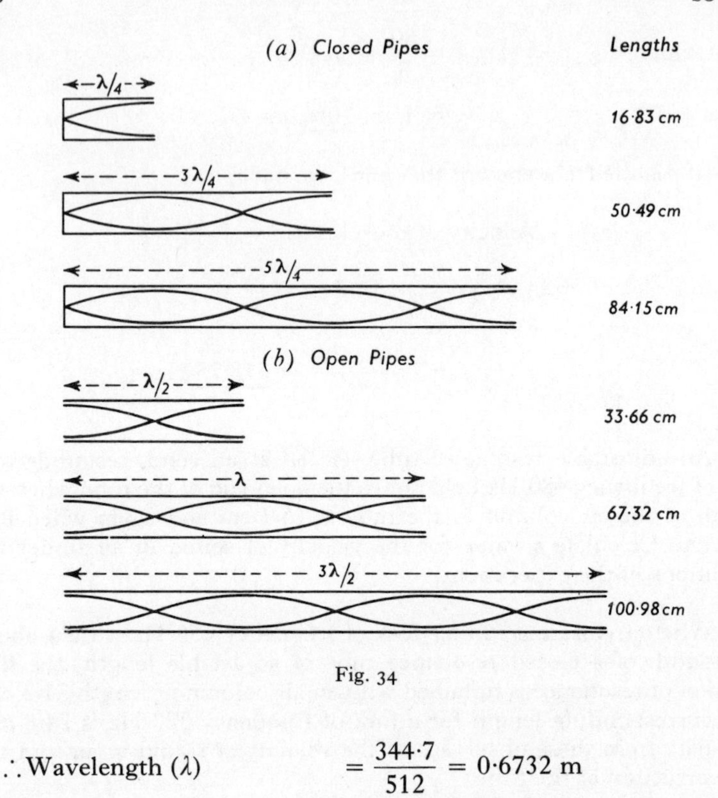

Fig. 34

$$\therefore \text{Wavelength } (\lambda) = \frac{344{\cdot}7}{512} = 0{\cdot}6732 \text{ m}$$

$$= \underline{67{\cdot}32} \text{ cm.}$$

The vibrations set up in the three shortest pipes are shown in the diagram, the lengths of the pipes being as indicated.

31 A rod, clamped at its mid point, is set up with a Kundt's tube containing air. When the rod is stroked longitudinally with a resined cloth it is found that the inter-node distance in the tube is 6·3 cm. If the air temperature is 15°C and the velocity of sound in air at 0°C is 331 m s^{-1}, calculate the frequency of the note emitted by the rod.

32 If the rod in the above question has a cross-sectional area of 1 cm^2, a length of 1 m, and mass 0·26 kg, obtain a value for Young's modulus of the material of the rod.

33 A Kundt's tube containing oxygen gas at a pressure of 765 mm of mercury and at a temperature of 17°C is adjusted to be in resonance with the exciting rod when it is found that the inter-node distance is 14·3 cm. If the note emitted by the rod is in unison with that sounded by an open organ pipe 15 cm long, obtain a value for the ratio of the specific heat of oxygen at constant pressure to that at constant volume. (Velocity of sound in air at 0°C = 331 m s^{-1}.)

Vibrations in strings

34 Prove that the velocity of propagation of transverse waves along a taut string is $\sqrt{\dfrac{T}{m}}$ where T is the tension in the string and m its linear density.

The mass of the vibrating length of a sonometer wire is 1·20 g, and it is found that a note of frequency 512 Hz is produced when the wire is sounding its second overtone. If the wire is in tension under a load of 10 kg, what is the vibrating length of the wire?

35 Describe how you would use a sonometer to determine the absolute frequency of a tuning-fork.

A certain tuning-fork is found to be in unison with 98 cm of a sonometer wire when the latter is set in transverse vibration. On shortening the wire by 1 cm it makes 4 beats per second with the fork. Assuming the wire to be in constant tension, calculate the frequency of the fork.

36 Describe how a sonometer could be used to determine the frequency of the a.c. mains supply. Show how you would calculate the result from the observations made.

37 A steel wire, having a mass of 0·2 g and a length of 1 m, is stretched between two smooth pegs under a load of 5 kg. The wire is now set in transverse vibration by plucking at its centre. What is the frequency of the note emitted by the wire?

38 Derive an expression for the velocity of propagation of longitudinal waves along a stretched wire.

A sonometer is fitted with wire having a diameter of 0·5 mm and which is in tension under a load of 5 kg. The wire is set in vibration (i) by stroking it with a resined cloth, and (ii) by plucking it, when it is found that the ratio of the fundamental frequencies of the notes emitted by the wire is 10:1. Obtain a value for Young's modulus of the wire from these observations.

39 Give an account of Melde's experiment and show how it can be used to demonstrate the laws of transverse vibration of a stretched string.

40 The relation between the tension (T) and the vibrating length (l) of a given sonometer wire when vibrating transversely with its fundamental frequency so as to be in unison with a given tuning-fork may be expressed by the relation Tl^p = const., where p is some numerical factor. Describe the details of an experimental method that would enable you to evaluate p in the above relation.

The length of a loaded sonometer wire is adjusted so as to give unison with a certain tuning-fork. An addition to the load is now made so as to increase the tension of the wire by 4 per cent, and when the same length of wire and the fork are sounded together 5 beats per second are heard. What is the frequency of the fcrk?

41 In Melde's experiment one end of a fine string is attached to a prong of a tuning-fork whilst the other end supports a load of 400 g. When the fork is set vibrating the string shows 3 loops. How must the load attached to the string be altered for the string to vibrate with 4 loops?

42 When the string attached to the tuning-fork in a Melde's experiment is in the plane of vibration of the prong's of the fork, four loops are observed when the string carries a given load. How many loops will the string show if now the plane of vibration of the fork is moved through a right angle? (Assume the load and length of string to be the same in each case.) Give reasons for your answer.

Worked example

43 *A sonometer wire* 40 *cm long and* 0·5 *mm diameter emits, when vibrating transversely, a fundamental note of the same frequency as that emitted by a siren whose disc, containing twenty-five holes, makes* 25 *revolutions per second. What is the tension in the wire if the density of its material is* $8{\cdot}9 \times 10^3$ *kg m*$^{-3}$?

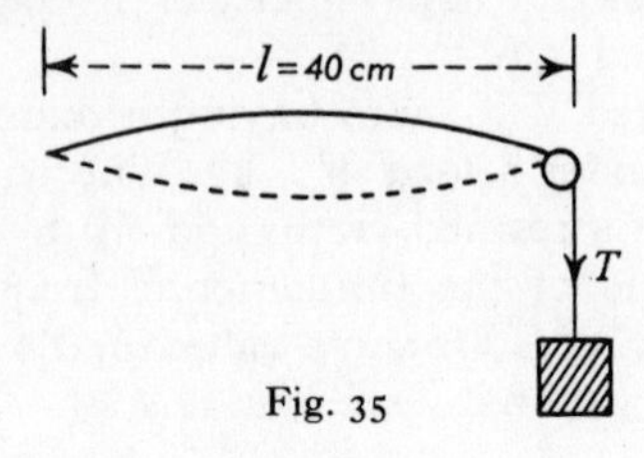

Fig. 35

The frequency (f) of the note emitted by a sonometer wire vibrating transversely is given by the expression

$$f = \frac{1}{\lambda}\sqrt{\frac{T}{m}}$$

where γ is the wavelength of the note emitted, T the tension of the wire, and m is the mass per unit length of the wire. For the fundamental note the wavelength is $2l$ (where l is the length of the vibrating wire), and if r and ρ are respectively the radius and density of the wire, we have, since $m = \pi r^2 \rho$,

$$f = \frac{1}{2l}\sqrt{\frac{T}{\pi r^2 \rho}}$$

Hence, inserting values,

$$f = \frac{1}{2 \times 0{\cdot}4}\sqrt{\frac{T}{\pi \times (2{\cdot}5 \times 10^{-4})^2 \times 8{\cdot}9 \times 10^3}}$$

from which

$$T = \underline{436{\cdot}8\ \text{N}}$$

44 The vibrating lengths of two wires, A and B, of the same material, are in the ratio 3 : 2. If the ratio of their tensions is 4 : 1, and the diameter of A is twice that of B, compare the frequencies of the fundamental notes emitted by the wires when in transverse vibration. If wire A has to be shortened by 10 cm to bring the two notes in unison, what were the original lengths of the wires?

45 A vibrating length of 50 cm of a sonometer wire, loaded with a 6-kilogramme mass, produces 5 beats per second when sounded simultaneously with a tuning-fork of frequency 256 Hz. If the wire has the lower frequency, find what change in (i) the tension, (ii) the length of the wire must be made to bring it in unison with the fork.

46 A sonometer wire is loaded with a heavy brass weight and it is found that a vibrating length of 80 cm gives unison with a certain tuning-fork. The weight is now completely immersed in water when it is found that the vibrating length of the wire has to be reduced to 75·1 cm to re-establish unison with the fork. Calculate the relative density of the brass weight.

Worked example

47 *A uniform sonometer wire* 1 *m long is divided into two segments by a bridge, the shorter segment being* 49·5 *cm long. When the two segments* *are simultaneously set in vibration* 6 *beats per second are heard. Calculate the frequencies of the two vibrations.*

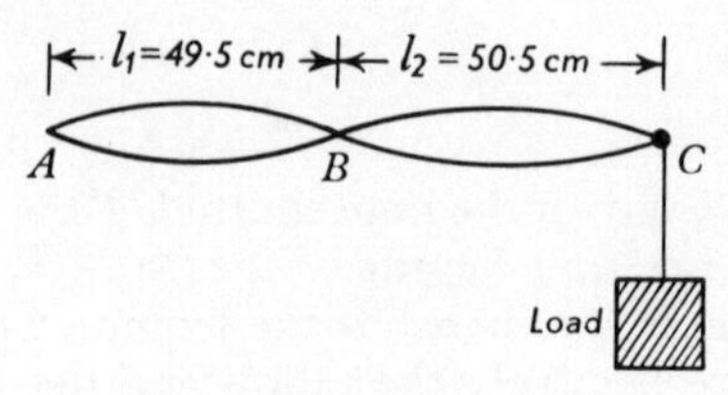

Fig. 36

The frequency (f) of transverse vibrations of a wire of length l sounding its fundamental note is given by the expression

$$f = \frac{1}{2l}\sqrt{\frac{T}{m}}$$

T being the tension in the wire, and m its mass per unit length. For the two sections AB (length l_1) and BC (length l_2) of the wire in question, T and m are constant. Hence if the frequencies of the notes emitted by these vibrating lengths are f_1 and f_2 respectively, we have

$$f_1 = \frac{\text{const.}}{l_1} \quad \text{and} \quad f_2 = \frac{\text{const.}}{l_2}$$

or
$$\frac{f_1}{f_2} = \frac{l_2}{l_1} = \frac{50{\cdot}5}{49{\cdot}5} \tag{1}$$

If the two notes are sounded together the frequency of the beats heard will be $f_1 - f_2$.

Thus
$$f_1 - f_2 = 6 \tag{2}$$

From (1)
$$f_1 = \frac{50{\cdot}5}{49{\cdot}5} f_2$$

and inserting this value for f_1 in equation (2) we have

$$\frac{50{\cdot}5}{49{\cdot}5} f_2 - f_2 = 6$$

from which
$$f_2 = \underline{297}\text{ Hz}$$
$$f_1 = \underline{303}\text{ Hz}$$

48 A given length of sonometer wire when loaded with 9·2 or 8·9 kg produces 3 beats per second on being sounded in conjunction with a certain tuning-fork. What is the fork frequency?

49 A set of wires of the same material and under constant tension are arranged in order of gradually diminishing length. The lengths of the wires are such that any one wire gives 5 beats per second when set in transverse vibration with a neighbouring wire. Four beats per second are heard when the 21st wire is vibrated with the 1st wire, and 6 beats per second are heard when it is vibrated with the 2nd wire. Explain these observations and calculate the frequency of the lowest note sounded by the wires.

50 A metre length of steel wire, of diameter 1 mm, is firmly clamped at one end and a load of 8 kg attached to the other end of the wire is observed to extend it by 0·5 mm. If the wire is now stroked with a resined cloth, calculate the frequency of the note emitted by the wire. (Density of steel $= 7{\cdot}8 \times 10^3$ kg m^{-3}.)

51 A rod of brass, 1 metre in length, is firmly clamped at points 25 cm from each end of the rod. Calculate the frequency of the note emitted by the rod when the section between the clamps is stroked with resined cloth, calculate the frequency of the note emitted by the wire. density of brass $= 8{\cdot}5 \times 10^3$ kg m^{-3}.)

Doppler effect

52 A source is emitting a note of constant frequency 512 Hz. What will be the apparent frequency of the note heard by an observer (*a*) if he is stationary and the source is approaching him with a velocity of 50 km hr^{-1}, (*b*) if the source is stationary and the observer is moving towards it with a velocity of 75 km hr^{-1} and (*c*) if both source and observer are moving towards each other with velocities of 50 and 75 km hr^{-1} respectively? Develop all formulae used. (Velocity of sound = 340 m s^{-1}.)

53 An engine, A, is moving away from a stationary engine, B, and meantime the two engines are continuously sounding their whistles, both of pitch 1000 Hz. If the driver of engine A hears beats of frequency 8 s^{-1}, what is the speed of his engine? Are the beats heard by the driver of engine B of the same frequency as those heard by the driver of engine A? (Velocity of sound in air = 349 m s^{-1}.)

Worked example

54 *An observer, standing by a railway track, notices that the pitch of an engine's whistle changes in the ratio 5:4 on passing him. What is the speed of the engine? (Velocity of sound in air = 340 m s^{-1}.)*

If f is the true pitch of the engine whistle, and v the speed of the engine, then, by the theory of the Doppler effect, the observed frequency (f_1) as the engine approaches is given by

$$f_1 = \left(\frac{c}{c-v}\right)f$$

and the observed frequency f_2 after the engine has passed is given by

$$f_2 = \left(\frac{c}{c+v}\right)f$$

where c is the velocity of sound in air.

Hence
$$\frac{f_1}{f_2} = \frac{\left(\frac{c}{c-v}\right)f}{\left(\frac{c}{c+v}\right)f} = \frac{c+v}{c-v} = \frac{5}{4}$$

from which
$$\frac{2v}{2c} = \frac{1}{9}$$

giving
$$v = \frac{c}{9} = \frac{340}{9}$$

$$= \underline{37{\cdot}8 \text{ m s}^{-1}} \quad \text{or} \quad \underline{136} \text{ km hr}^{-1}$$

55 A whirling whistle of frequency 500 Hz is revolved at the rate of 400 revolutions per minute on the end of a string of length 1·2 m. What is the range of frequencies heard by an observer standing some distance away in the plane of rotation of the whistle? (Velocity of sound in air = 340 m s^{-1}.)

56 Discuss the effect of wind on the Doppler effect in sound.

Two stationary observers, A and B, situated at places some distance from each other along a straight railway track, are listening to a train's whistle which is being continuously sounded as a train travels at a speed of 90 km hr^{-1} from A to B. If the whistle has a true frequency of 1000 Hz, and if the component of the wind velocity along AB is 30 km hr^{-1} in the direction from A, calculate the apparent frequencies of the notes heard at A and B. Velocity of sound = 340 m s^{-1}.

57 An aircraft, flying at a speed of 600 km hr^{-1} on a horizontal track at a height of 300 m, is continuously sounding a note of frequency 1000 Hz. Find the change in the apparent frequency of the note as heard by an observer on the ground during an interval of one second as it passes overhead.

58 A car, A, travelling at 90 km hr^{-1}, is approaching a side junction and is continuously sounding its horn (frequency 300 Hz), whilst another car, B, travelling at 45 km hr^{-1} is approaching the main road. Assuming both roads to be straight and mutually at right angles, calculate the frequency of the note heard by the driver of car B when the line joining the two cars makes an angle of 45° with the roads. (Velocity of sound = 340 m s^{-1}.)

The doppler effect in light

59 Calculate the broadening, due to random thermal motion, of the line of wavelength 5890 Å emitted by a sodium lamp running at a temperature of 300°C. (Gas constant = 8·31 J mol^{-} K^{-1}; relative atomic mass of sodium = 23; velocity of light = 3×10^{8} m s^{-1}; 1 Å = 10^{-10} m.)

60 The line of wavelength 4340 Å in the spectrum of a given star is found to be displaced towards the red end of the spectrum by 1 Å in reference to the line position in a comparison spectrum of a laboratory source. What is the radial velocity of the star? (Velocity of light = 3×10^{8} m s^{-1}.)

61 When examined under laboratory conditions the wavelength of a given hydrogen line is found to be 4861·37 Å. When observed in the spectrum of a given star this line appears as a broadened band with a wavelength spread of from 4861·64 Å to 4861·88 Å. Account for this and conclude what you can concerning the motion of the star. (Velocity of light = 3×10^{8} m s^{-1}.)

62 Comparison of the photographs of the sun's spectrum taken with the slit of a spectrograph directed towards opposite limbs of the sun's equator reveal a relative displacement of the *F* line in the two spectra of 0·065 Å. If the line has a wavelength of 4861 Å when produced by a laboratory source, and the period of axial rotation of the sun is 25 days 9 hours, calculate the sun's equatorial diameter. (Velocity of light = 3×10^{8} m s^{-1}.)

ELECTROSTATICS

Some formulae and units	*units*
Permittivity of free space $= \varepsilon_0 = \dfrac{10^{-9}}{36\pi}$ or $8{\cdot}84^{-12}$	$\mathrm{F\,m^{-1}}$
Relative permittivity $= \varepsilon_r = 1$ for vacuum (or air)	
Absolute permittivity $= \varepsilon = \varepsilon_r \varepsilon_0$	$\mathrm{F\,m^{-1}}$
Electric field strength (intensity) $= E$ $=$ potential gradient $\left(-\dfrac{dV}{dx}\right)$	$\mathrm{N\,C^{-1}}$ or $\mathrm{V\,m^{-1}}$
Quantity of charge $= Q$	C
Force between point charges $= F = \dfrac{Q_1 Q_2}{4\pi\varepsilon_0\varepsilon_r x^2}$	N
Force on charge in electric field $= QE$	N
Electric charge density (surface) $= \sigma$	$\mathrm{C\,m^{-2}}$
Electric field at surface of a conductor $= \dfrac{\sigma}{\varepsilon}$	$\mathrm{V\,m^{-1}}$
Force (per unit area) of charged conductor $= \dfrac{\sigma^2}{2\varepsilon}$	$\mathrm{N\,m^{-2}}$
Pull between charged plates $= \dfrac{\varepsilon A V^2}{2d^2}$	N
Potential at a point in field of a charged body $= \dfrac{Q}{4\pi\varepsilon x}$	V
Capacitance $C = \left(\dfrac{Q}{V}\right)$	F
Capacitance of an isolated sphere $= 4\pi\varepsilon r$	F
Capacitance of a parallel plate capacitor $= \dfrac{\varepsilon A}{d}$	F
Energy of a charged conductor $= \frac{1}{2}CV^2$ (or $\frac{1}{2}QV$)	J

Some useful prefixes:

micro (μ) 10^{-6}
nano (n) 10^{-9}
pico (p) 10^{-12}

Forces between charges. Potential

1 State the law of force between two point charges.

Two small conducting spheres (of radii 1 and 2 cm respectively) are initially in contact and are given a certain quantity of charge. The spheres are then separated (without losing their acquired charge) until their centres are 20 cm apart when it is found that the force of repulsion between them is 5μN. What was the magnitude of the original charge given to the spheres? Find also the point of zero intensity in the combined field of the charged spheres. (Permittivity of free space

$$= \frac{10^{-9}}{36\pi}\,\mathrm{F\,m^{-1}}.)$$

Worked example

2 *Two small identical conducting spheres each of mass* 1 *g are suspended by silk threads* 20 *cm long from the same point. On being charged the spheres, which were originally in contact, take up positions with their centres* 5 *cm apart. What was the original charge given to the spheres? (Acceleration of gravity* = 9·8 *m* s^{-2}, *and the permittivity of free space* (ε_0) is $\frac{10^{-9}}{36\pi}$ *F* m^{-1}.)

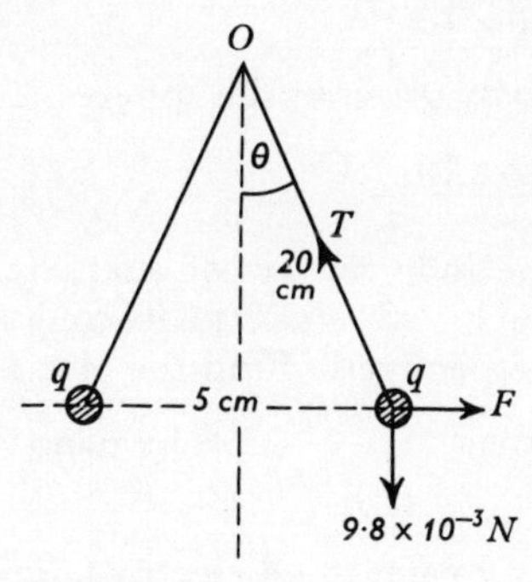

Fig. 37

Since the spheres are identical in size, they will take equal quantities of charge when originally charged. Let this quantity be q. Then the force of repulsion keeping them apart in the equilibrium position is

$$\frac{q^2}{4\pi\varepsilon_0(0{\cdot}05)^2}\quad \mathrm{N}.$$

The diagram shows the forces on one of the spheres in this position. These forces are: the force of repulsion $\left(\frac{q^2}{4\pi\varepsilon_0(0{\cdot}05)^2}\right)$N , the force of gravity ($9{\cdot}8 \times 10^{-3}$ N) and the tension T in the thread acting in the directions shown. If the thread makes an angle θ with the vertical through 0, we have, resolving along the vertical and horizontal through the mass centre of the sphere,

$$T \sin \theta = \frac{q^2}{4\pi\varepsilon_0(0{\cdot}05)^2} \text{ (horizontal component forces)} \qquad (1)$$

$$T \cos \theta = 9{\cdot}8 \times 10^{-3} \text{ (vertical component forces)} \qquad (2)$$

By division these equations give

$$\tan \theta = \frac{\dfrac{q^2}{4\pi\varepsilon_0(0{\cdot}05)^2}}{9{\cdot}8 \times 10^{-3}} = \frac{36}{9{\cdot}8} \times 10^{14}\, q^2$$

Now $\quad \tan \theta = \dfrac{2{\cdot}5}{\sqrt{20^2 - (2{\cdot}5)^2}} = 0{\cdot}126$

Thus $\quad q^2 = \dfrac{0{\cdot}126 \times 9{\cdot}8}{36 \times 10^{14}}$

from which $\quad \underline{q = 2{\cdot}06 \times 10^{-9}\text{ C}}$

or, total charge originally given to the spheres

$$= \underline{4{\cdot}12 \times 10^{-9}\text{ C}}$$

3 A, B, C, D are the four corners of a square of side 10 cm. Point charges of magnitudes 1, -2, and 3 nano-coulomb are situated at the corners A, B and C respectively. Find the electrostatic potential at D. (Permittivity of free space $= \dfrac{10^{-9}}{36\pi}$, prefix nano $= 10^{-9}$.)

4 Define *potential at a point* in an electrostatic field.

An isolated conducting sphere of radius 5 cm is given a positive charge of 4×10^{-9} C. Calculate the potential at points 2, 5 and 15 cm from the centre of the sphere.

5 What do you understand by an 'equi-potential surface'?

Two spheres, A (of 4 cm radius) and B (of 2 cm radius) have charges of $+10 \times 10^{-9}$ C and -4×10^{-9} C respectively and are positioned with their centres 30 cm apart. Find the potentials at points on the line joining their centres which are distant 5, 15 and 25 cm from A's centre. Find also the point of zero potential in the conjoint field and draw the general shape of the equipotential contours of the field.

6 With the aid of a labelled diagram describe the assembly of a Van de Graaff generator and explain the physical principles involved in its action.

The high voltage sphere of such a generator has a radius of 0·5 m. To what voltage can it be raised under normal atmospheric conditions given that electrical breakdown occurs when the electrical intensity exceeds 3×10^6 V m^{-1} for such conditions.

7 Given that under average weather conditions there is an electrical intensity gradient of 100 volt per metre near the surface of the Earth, calculate the corresponding electrical charge density on the Earth's surface.

Capacitors. Energy of charge

8 Upon what factors does the capacitance of a parallel plate capacitor depend?

A parallel plate air capacitor is charged to a potential difference of 200 volt. It is then connected with its terminals in parallel with an uncharged capacitor of similar dimensions but having ebonite as its dielectric medium. The potential of the combination is found to fall to 50 volt. What is the relative permittivity of the ebonite?

9 You are provided with three identical capacitors each of capacitance 2 μF. What is (*a*) the maximum, (*b*) the minimum capacitance you could obtain with these capacitors? What arrangement of the capacitors would give a resultant capacitance of 3 μF?

10 A parallel plate capacitor is placed with its plates vertical inside a large beaker and with one of its plates permanently earthed. The capacitor is then charged by connecting a 100 V battery across it. The battery connections are now removed and oil of relative permittivity 5 is poured into the beaker so as to submerge the bottom quarter of the capacitor system. What is then the potential difference across the capacitor plates?

11 Derive an expression for the effective capacitance of three capacitors arranged in series.

Uncharged capacitors of 2 μF, 3 μF and 6 μF capacitance values are

connected in series across 120 V d.c. mains. Find
(*a*) the quantity of charge drawn from the mains,
(*b*) the potential difference across each capacitor,
(*c*) the energy stored in the 3 μF capacitor.

12 The plates of a parallel-plate air capacitor, with a plate area of 100 cm^2, are connected across a battery with a terminal p.d. of 100 volt. What is the energy of the charged capacitor if the distance between the plates is 1 cm? How is the energy changed if the distance between the plates is increased to 2 cm (i) with the capacitor plates remaining connected to the battery, (ii) if the plates are disconnected before being moved?

Worked example

13 *Two capacitors of capacitance 3 and 2 μF respectively are connected in series, the free plate of the smaller being connected to earth. If the free plate of the larger is charged to a potential of 200 volt, determine the potential difference across the smaller capacitor and the energy stored in it.*

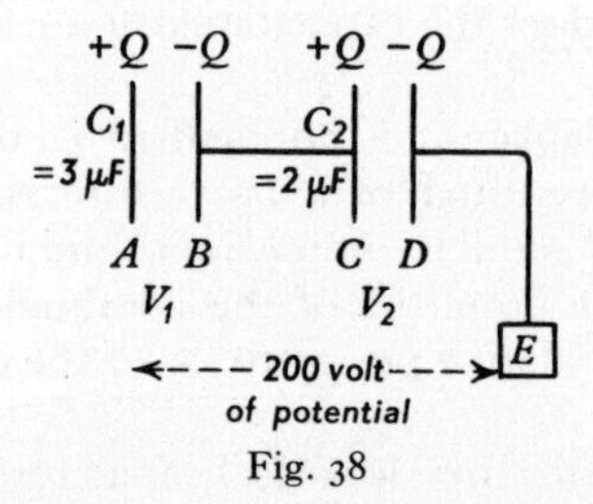

Fig. 38

The diagram shows the arrangement of the capacitors the plates of which are labelled A, B and C, D. Plate A of the first capacitor is raised to a potential of 200 volt above the earth, and if the potential difference across the plates AB is V_1 and that across the plates CD is V_2, we have

$$V_1 + V_2 = 200 \tag{1}$$

Now if the charge on plate A is $+Q$ units, there will be an induction displacement of charge of $-Q$ units on plate B, $+Q$ units on plate C, etc. so that both capacitors carry the same charge. Then, for the first capacitor

$$Q = C_1 V_1 = 3 \times 10^{-6} \; V_1$$

and for the second

$$Q = C_2 V_2 = 2 \times 10^{-6} \; V_2$$

Hence $3\times10^{-6}\,V_1 = 2\times10^{-6}\,V_2$

or $V_1 = \frac{2}{3}V_2$ (2)

Inserting this value of V_1 in equation (1), we have,

$$\frac{2}{3}V_2+V_2 = 200$$

from which $V_2 = \underline{120}$ volt

The electrostatic energy of the second capacitor

$$= \tfrac{1}{2}C_2V_2^2$$
$$= \tfrac{1}{2}\times 2\times10^{-6}\times(120)^2$$
$$= \underline{1{\cdot}44\times10^{-2}\ \text{J}}$$

14 Derive an expression for the energy stored in a charged capacitor.

Two capacitors, of 2 and 3 μF, are joined (*a*) in series, (*b*) in parallel and then connected across a battery of 100 V. What is the energy stored in the 2 μF capacitor in each case?

15 Show that when two capacitors share their charges there is always a resulting energy loss. Account for this loss.

Two capacitors, of capacitances 2 and 5 μF each receive a charge of 100 μC. They are then connected together. Calculate,
(*a*) the common potential of the capacitors,
(*b*) the energy loss after connection.

16 A variable air capacitor, of maximum capacitance 5×10^{-9} F, is charged by connecting it up to a 500 V supply. It is then disconnected from the supply, and its capacitance varied until a spark discharge occurs between the plates. If this occurs when the potential difference across the air gap between the plates is 2500 volt, calculate the inter-plate capacitance at breakdown and the energy of the discharge. Compare this energy with the maximum energy initially stored across the 500 volt supply and comment on the result.

17 Three capacitors with capacitances of 2, 3 and 5 μF respectively are connected in series and a d.c. source of 1000 V applied across them. What is the charge and potential difference of each capacitor? How does the energy stored in this arrangement of the capacitors compare with that stored in them if they had been connected in parallel across the source of voltage?

18 What do you understand by the terms *absolute* and *relative permittivity*?

The charged plate of a parallel plate air capacitor is connected to a sensitive electroscope which records a deflection of 50 divisions. If the air space between the plates is now completely filled with another insulating medium, the deflection of the electroscope falls by 30 divisions. What is the relative permittivity of the new medium?

19 A parallel plate air capacitor has its plates situate 3 cm apart. One of the plates is earthed and the other connected to a sensitive electroscope the deflection of which is observed when the capacitor is charged. A slab of glass 1·5 cm thick, and large enough to cover the plate area of the capacitor, is now placed between the capacitor plates when it is noticed that the deflection of the electroscope falls. Explain this and calculate a value for the relative permittivity of the glass if, to restore the deflection, it is necessary to increase the residual air gap between the plates by 1·25 cm.

20 A parallel plate air capacitor with plate area 0·03 m^2 is charged by connecting it to a d.c. source providing a constant p.d. of 33 V across its plates. After each charging the positive plate is connected to a sensitive d.c. amplifier calibrated to read quantities of charge, and the following table gives the value of the charge (q) on the capacitor as the distance (d) between its plates is successively varied.

q	11·0	8·0	6·6	5·7	5·0	4·7	4·4	$\times 10^{-10}$ C
d	0·010	0·015	0·020	0·025	0·030	0·035	0·040	m

Plot a suitable graph of these values, and from the graph determine
(*a*) a value for the *electric space constant* ε_0,
(*b*) the stray capacitance associated with the capacitor during the exercise.

21 A capacitor A having a capacitance of 4 μF is charged from a cell and when discharged through a ballistic galvanometer produces a scale deflection of 30 divisions. The capacitor is again charged up from the same cell and, through a suitable switch, a second capacitor B is successively charged from A and subsequently discharged. After five such steps of drawing charge from A, the residual charge on it produces a scale deflection of 10 divisions when discharged through the ballistic galvanometer.

Sketch a suitable circuit for this exercise and calculate the capacitance of B.

22 A parallel plate capacitor of capacitance C_1 is charged so that the potential difference across its plates is V_0. The plate separation of the capacitor is now increased to n times its original value when it is connected to an uncharged capacitor of capacitance C_2. What must be the value of C_2 for the potential difference across the capacitors still to be V_0?

23 A d.c. pulse generator of calibrated frequency, a reed (or vibrator) switch, a battery of known e.m.f., and a sensitive milliammeter are provided to make a determination of the capacitance of a given capacitor. Draw a diagram of the necessary circuit arrangements and calculate the appropriate capacitance value if the battery used has an e.m.f. of 25 volt and a steady current of 8·0 mA is recorded by the ammeter when the reed makes 200 vibrations per second.

24 The diagram shows three thin plane-parallel metal plates arranged with air separations as marked. Plates A and C are connected together

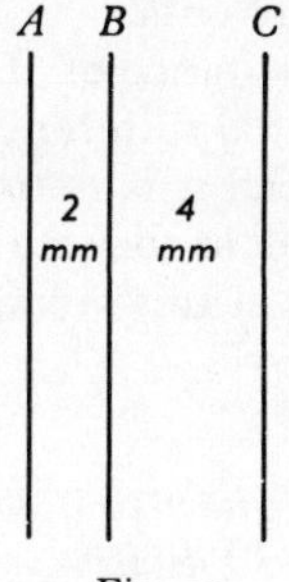

Fig. 39

and plate B raised to a potential of 50 V above A and C. A and C are now disconnected and C is then raised to a potential of 25 V above A. What is now the potential difference between B and A? Ignore edge effects of the plates.

25 A 30 V battery is used to charge in turn three capacitors (AB, CD, EF) of capacitances 2, 3 and 5 μF respectively, the positively charged plates being A, C and E. The capacitors are then connected in a ring formation by connecting B to C, D to E and F to A. What are now the charges on A, C and E, and how does the potential vary round the circuit?

26 A slab of glass of thickness 2 cm and relative permittivity 5 is laid flat on the bench and a plane metal sheet, of area 2×10^{-2} m^2, placed over its upper surface. A similar metal sheet is supported in an insulating stand symmetrically above the first metal sheet and 1 cm distant from it, and a potential difference of 1500 volt established between the sheets. Calculate the quantity of charge stored in the arrangement if (*a*) the lower sheet is earthed, (*b*) the upper sheet is earthed.

27 Derive expressions for the capacitance of a spherical shell capacitor (*a*) when the outer shell is earthed, (*b*) when the inner shell is earthed.

Three insulated concentric hollow spheres A, B and C have radii of 2, 3 and 5 cm and carry charges of 3, 4·5 and 6 nC respectively. What are the potentials of the spheres? If the middle sphere is now earthed, what effect would this have on the potentials of the outer two spheres? Also, what is now the charge on the middle sphere?

Electrostatic instruments. Forces on charged bodies

28 Describe the use of an electrometer (or electroscope) for the measurement of ionization currents.

An electroscope has a capacitance of 10^{-11} F and gives a deflection of 30 divisions when a potential difference of 200 volt is applied to it. When charged to this potential it is noticed that the deflection of the electroscope commences to fall at the rate of half a division per minute. Calculate the leakage current at this voltage.

Worked example

29 *A charge of electricity is given to a gold leaf electroscope the leaves of which record a deflection of* 30 *divisions. An uncharged metal sphere of radius* 5 *cm is now placed in contact with the cap of the electroscope when the divergence of the leaves falls to* 10 *divisions. What is the capacitance of the electroscope?*

Let C_E be the capacitance of the electroscope and let the charge given to it be Q. Then

$$Q = C_E V_1 \tag{1}$$

V_1 being the initial potential recorded by the electroscope.

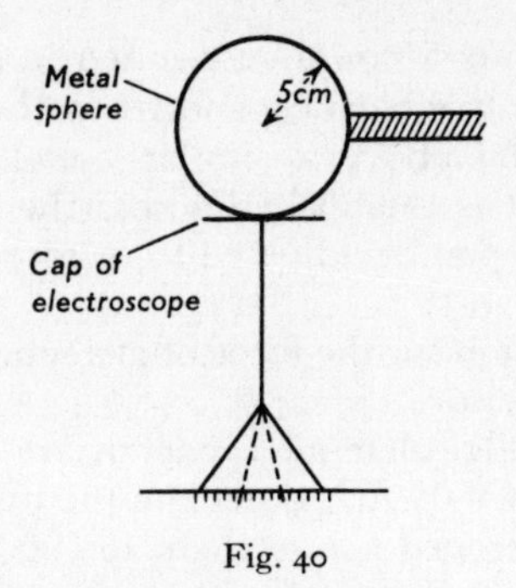

Fig. 40

Now when the sphere is placed in contact with the electroscope, the total capacitance of the system is $C_E + C_S$ (C_S being the capacitance of the metal sphere $= 4\pi\varepsilon_0 r$), and since the charge is unchanged, we now have

$$Q = (C_E + C_S)\,V_2 \tag{2}$$

V_2 being the new potential recorded by the electroscope. Hence from (1) and (2)

$$(C_E + C_S)\,V_2 = C_E V_1$$

or

$$1 + \frac{C_S}{C_E} = \frac{V_1}{V_2} = \frac{30}{10}$$

(since the deflection is proportional to the potential)
from which

$$C_E = \frac{C_S}{2} = \frac{4\pi\varepsilon_0 \times 0{\cdot}05}{2}$$

$$= \frac{4\pi}{2} \times \frac{10^{-9}}{36\pi} \times 0{\cdot}05 \qquad \left(\varepsilon_0 = \frac{10^{-9}}{36\pi}\ \mathrm{F\,m^{-1}}\right)$$

$$= \underline{2{\cdot}78\ pF}.$$

30 Describe the quadrant electrometer and explain its use in measuring small statical potential differences.

The potential difference between the plates of a charged capacitor is found to fall to half its value within a time of 12 seconds when the plates of the capacitor are joined through a 2-megohm resistor. What is the capacitance of the capacitor?

31 Describe the Kelvin attracted disc electrometer.

A metal disc of radius 5 cm is suspended in a horizontal position from one arm of a sensitive balance, and when counterpoised is 0·5 cm symmetrically vertically above a similar metal disc. If a potential difference of 1000 volt is established across the plates how must the weights in the balance pan be adjusted to restore counterpoise?

32 Obtain an expression for the force of attraction between the plates of a parallel plate capacitor.

The plates of a parallel plate air capacitor are set 0·01 m apart and have an overlap area of 0·05 m^2. Calculate the pull between the plates when the voltage difference across them is 100 V and if the electric space constant has a value of $8{\cdot}84 \times 10^{-12}$ F m^{-1}.

33 The parallel plate air capacitor of the above problem is disconnected from the source after being charged in the manner indicated and a second parallel plate air capacitor is connected in parallel with it. It is then observed that the pull between the plates of the first capacitor is reduced to half its former value. What is the capacitance of the second capacitor?

34 If, now, the inter-plate space of the second capacitor of problem 33 is completely filled with another insulating medium, and, without altering the distance between the plates, the pull between the first capacitor plates is reduced to one quarter of its original value, what is the relative permittivity of the new medium?

35 Derive an expression for the mechanical force per unit area of a charged conductor.

A spherical soap bubble of radius 1 cm is blown in an evacuated enclosure. Find what charge must be given to the bubble to double its radius. (Take the surface tension of soap solution as 3×10^{-2} N m^{-1}, and ignore evaporation effects from the surface of the bubble.)

36 A small particle carrying a negative charge of $1{\cdot}6 \times 10^{-19}$ C is suspended in equilibrium between two horizontal metal plates 4 cm apart having a potential difference of 3000 volt across them. What is the mass of the particle? (Take the acceleration of gravity as 9·8 m s^{-2}.)

CURRENT ELECTRICITY

Ohm's law and resistance. Kirchoff's laws

1 State *Ohm's law* and describe how you would give experimental verification of it.

A series circuit, consisting of two cells, A and B, and a high resistance galvanometer, is set up and the galvanometer deflection is observed to be 31·5 divisions. On reversing cell B in the circuit, the galvanometer deflection becomes 4·5 divisions. Compare the e.m.f.s of the two cells.

2 Describe some method of finding the internal resistance of a cell.

A given cell is connected in circuit with a variable resistance whilst, at the same time, a high resistance voltmeter is connected across the terminals of the cell. The voltmeter reads 1·30 volt when the variable resistance is fixed at 13 ohm, and again 1·20 volt when this resistance is at 8 ohm. Calculate the e.m.f. and the internal resistance of the cell from this data.

3 A certain cell has an e.m.f. of 1·45 volt and an internal resistance of 4·5 ohm. Calculate the approximate percentage error of taking the reading of a voltmeter, having a resistance of 100 ohm, as the e.m.f. of this cell when the voltmeter is directly connected across its terminals. (The voltmeter is to be assumed correctly graduated.)

Worked example

4 *A cell of constant e.m.f. and negligible resistance is connected in series with a voltmeter and a variable resistance R. When R has a value of* 10 *ohm the voltmeter reads* 2·0 *volt, but when R is increased to* 100 *ohm the voltmeter reading falls to* 1·4 *volt. Find the resistance of the voltmeter and the e.m.f. of the cell.*

Let E = e.m.f. of the cell and R' = resistance of the voltmeter. Then the current I in the circuit is given by

$$I = \frac{E}{R'+R}$$

and thus the potential drop across the voltmeter

$$= IR'$$

$$= \frac{ER'}{R'+R} \text{ volt}$$

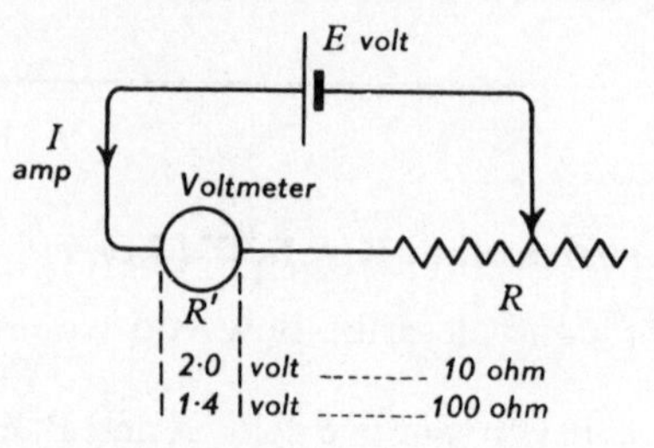

Fig. 41

Hence, inserting values, we have

$$\frac{ER'}{R'+10} = 2{\cdot}0 \qquad (1)$$

and

$$\frac{ER'}{R'+100} = 1{\cdot}4 \qquad (2)$$

These equations give, on division,

$$\frac{R'+100}{R'+10} = \frac{2{\cdot}0}{1{\cdot}4}$$

from which $R' = \underline{200}$ ohm

Inserting this value for the voltmeter resistance in equation (i), we have

$$\frac{200E}{200+10} = 2$$

giving $E = \underline{2{\cdot}1}$ volt

5 Derive the condition for which a given number of cells will supply a maximum current in an external resistor of fixed value.

Thirty cells, each of e.m.f. 2 volt and internal resistance 0·5 ohm, are used to send a current through an external resistor of value 5 ohm. What is this current if the cells are grouped (*a*) in series, (*b*) in parallel? What grouping of the cells will give a maximum current? What is the value of this current?

6 A galvanometer, an unknown resistor R, and a known resistor of 750 ohm are connected in series across an accumulator of negligible internal resistance. The deflection recorded by the galvanometer is 20 divisions, but this increases to 30 divisions on cutting out the 750-ohm resistor, and then to 45 divisions when both the 750-ohm resistor and the unknown resistor are out of circuit. Find the resistance of the galvanometer and the value of the unknown resistor.

7 Establish the formula for the effective resistance of a set of resistors in parallel.

Three wires, having resistances of 2, 5, and 10 ohm respectively, are joined in parallel and connected across a 12-volt battery. If the current through the 5-ohm wire is found to be 1·5 ampere, what is the resistance of the battery?

8 Distinguish clearly between electromotive force and potential difference.

A cell, A, a direct reading galvanometer and a plug resistance box are connected in series and the resistance box adjusted until the galvanometer shows a deflection of 25 divisions. On taking a further 15 ohm out of the box the deflection of the galvanometer falls to 20 divisions. The cell, A, is now replaced by another cell, B, in the circuit, and the resistance box is adjusted so that the galvanometer again shows a deflection of 25 divisions. Now, however, the box resistance has to be increased by a further 12 ohm in order to reduce the galvanometer deflection to 20 divisions. Compare the e.m.f.s of the two cells and explain your method.

9 Define *resistivity*.

Taking the resistivity of copper as $1{\cdot}6 \times 10^{-8}$ ohm-metre units, calculate the resistance of a cubic centimetre of copper (*a*) when in the form of a wire of diameter 0·02 cm, (*b*) when in the form of a sheet 2·5 mm thick, the current passing through the sheet perpendicularly to its faces.

10 Two wires, A and B, have lengths which are in the ratio of 4 : 5, diameters which are in the ratio 2 : 1, and are made of materials with resistivities in the ratio 3 : 2. If the wires are arranged in parallel, and a current of 1 ampere enters the arrangement, find the current in each branch.

11 Define the *temperature coefficient of resistance* of a wire.

What length of copper wire of diameter 0·1 mm is required to make a coil with resistance 0·5 ohm? If, on passing a current of 2 ampere, the temperature of the coil rises by 10°C, what error would arise by taking the potential drop across the coil as 1 volt? (Resistivity of copper $= 1{\cdot}6 \times 10^{-8}$ ohm-metre units. Temperature coefficient of resistance of copper $= 0{\cdot}0043\ K^{-1}$.)

12 How does the resistance of the following vary with temperature: (*a*) copper, (*b*) Eureka, (*c*) copper sulphate solution, (*d*) carbon?

A piece of iron wire has a resistance of 4·02 ohm at 100°C. What is the temperature coefficient of resistance of iron? At what temperature would the resistance of the wire be 3·00 ohm?

Worked example

13 *Two cells, E_1 of e.m.f. 2·0 volt and internal resistance 1·5 ohm, and E_2 of e.m.f. 1·5 volt and internal resistance 1·0 ohm, are joined in parallel with like poles together. Calculate the current that would pass through a 5 ohm resistor joined in parallel with the cells.*

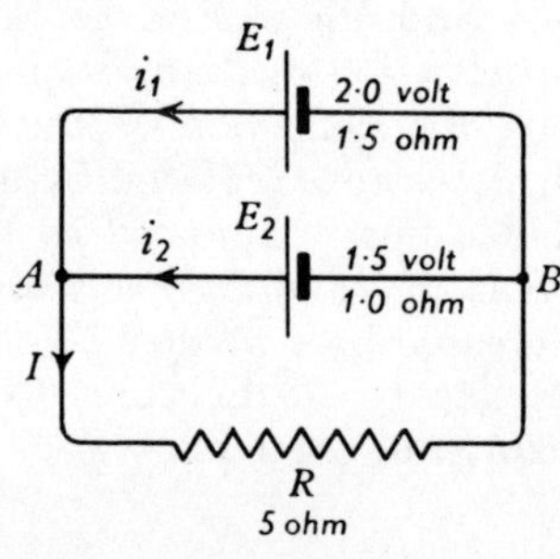

Fig. 42

The circuit arrangements are shown in the diagram, A being the common positive connection for the cells and B being the common negative connection. Let the currents in the various branches be i_1, i_2 and I as indicated, where $I = i_1 + i_2$ (since, by Kirchoff's 1st law, the algebraic sum of the currents at a junction is zero). Now considering the closed circuit E_1ARB, the net e.m.f. acting is 2·0 volt and this must equal the sum of the potential drops in the various parts of the circuit (by Kirchoff's 2nd law). Hence,

$$1{\cdot}5i_1 + 5I = 2$$

i.e.

$$1{\cdot}5i_1 + 5(i_1 + i_2) = 2$$

or

$$6{\cdot}5i_1 + 5i_2 = 2 \qquad (1)$$

By a similar consideration of the closed circuit E_2ARB we get

$$i_2 + 5I = 1{\cdot}5$$

i.e.

$$i_2 + 5(i_1 + i_2) = 1{\cdot}5$$

or

$$5i_1 + 6i_2 = 1{\cdot}5 \qquad (2)$$

The simultaneous equations (1) and (2) on solution give

$$i_1 = \tfrac{9}{28} \text{ ampere}$$

and
$$i_2 = -\tfrac{1}{56} \text{ ampere}$$

(The negative sign here indicates that, in fact, the current in this branch passes from B to A and *not* from A to B as indicated.)

Thus the current I through the 5-ohm resistor

$$= \tfrac{9}{28} - \tfrac{1}{56}$$
$$= \tfrac{17}{56} \text{ ampere}$$

14 Three cells of e.m.f.s 2·0, 1·5 and 1·1 volt respectively, and with respective internal resistances of 0·5, 2 and 5 ohm, are connected in parallel with like poles together. Calculate the potential drop across the terminals of this arrangement when an external resistance of 10 ohm is joined in circuit.

15 Derive the condition for balance of a Wheatstone bridge network.

The accompanying diagram shows such a network in which R, R are two equal fixed resistors, X a resistor of unknown value and S a continuously variable standard resistor. G is a galvanometer of resistance 250 ohm which has a full scale deflection of 1 mA and is capable

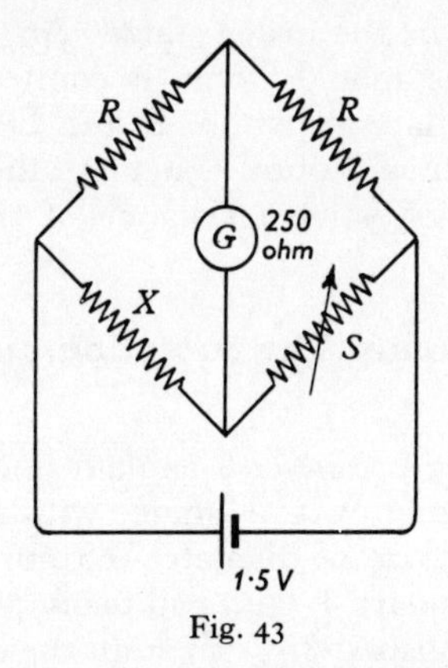

Fig. 43

of being read to a scale accuracy corresponding to 10μA. Calculate the value of R so that whatever the values of X or S, the galvanometer is not overloaded.

Find also the accuracy to which the resistance of a 5-ohm resistor can be measured using this arrangement.

16 A cube is composed of 12 equal wires each of resistance 2 ohm. If a current of 10 ampere enters at one corner of this mesh and leaves at the opposite corner, what is the potential drop across the mesh?

17 The diagram shows an electrical network in which only some of the currents and resistances are known—as indicated. Find (i) the

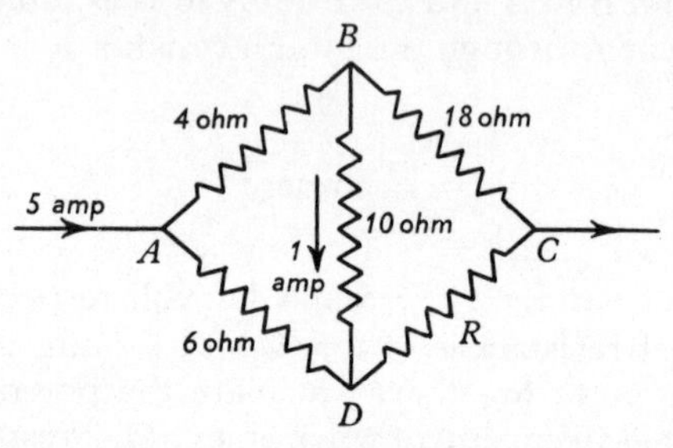

Fig. 44

value of R, (ii) the current along BC, (iii) the potential drop between the points A and C.

18 State Kirchoff's laws for electrical networks.

Four resistors of values 1, 3, 4 and 2 ohm are connected together to form a quadrilateral arrangement ABCD with the resistors in the sides AB, BC, CD and DA in the order stated. An accumulator of e.m.f. 2 volt and internal resistance 0·5 ohm is connected across the points A and C of the quadrilateral, whilst B and D are joined through a galvanometer of resistance 5 ohm. Calculate the current through the galvanometer and the effective resistance of the mesh between the points A and C.

Electrical measurements. The potentiometer and Wheatstone bridge

19 Describe the metre bridge and explain how you would use it to obtain the value of an unknown resistance with maximum accuracy.

A 120-cm length of wire of diameter 0·5 mm is placed in one gap of a metre bridge, a standard 1-ohm coil being placed in the other gap. A balance point is established 57·7 cm from the end of the wire corresponding to the 1-ohm coil. Calculate the resistivity of the wire from this data.

20 Resistors P, Q, R and S, arranged in clockwise order, comprise a Wheatstone bridge network, P and Q being the 'ratio' resistors.

R has a value of 20 ohm and it is found that a balance is obtained by shunting R with a resistance of 80 ohm. P and Q are now interchanged, when it is found that the shunt resistance across R has to be reduced to 5 ohm in order to restore balance. Find the value of S and the ratio $P:Q$.

21 Explain fully the significance and meaning of 'end corrections' in a bridge wire.

With resistances of 3 and 2 ohm in the gaps of a metre bridge a balance is obtained when the sliding contact is positioned at the 59·5 cm mark on the wire. On interchanging the 3 and 2 ohm resistances, the new balance is obtained at the 39·0 cm mark. Find the 'end corrections' of the wire.

22 Establish from first principles the relation between the four resistances of a balanced Wheatstone bridge.

Two resistors, X and Y, are combined and inserted in one gap of a metre bridge, a 10-ohm resistor being inserted in the other gap. When X and Y are in series a balance is obtained 24·5 cm from the end corresponding to the 10-ohm resistor, and when they are in parallel the balance point is 61·0 cm from the same end. What are the values of X and Y?

23 Give the theory of the method of comparing the resistances of two nearly equal resistors using the Carey-Foster modification of the Wheatstone bridge.

Show that balance positions using such a bridge in the above experiment are only possible provided that the resistances of both resistors are less than the resistance of the bridge wire.

24 Explain the principle of the potentiometer and describe how you would use a potentiometer to compare the e.m.f.s of two cells.

With a Leclanché and a Daniell cell connected in series so that their e.m.f.s act in conjunction, a balance is obtained against 87·5 cm of a potentiometer wire. On reversing the Daniell cell, the cells only require 14·0 cm of the potentiometer wire to secure a balance. Given that the e.m.f. of the Daniell cell is 1·05 volt, find that of the Leclanché cell.

Worked example

25 *A Leclanché cell, on open circuit, gives a balance point against* 82·1 *cm of a potentiometer wire. On now connecting a resistance of* 10 *ohm across the terminals of the cell, a balance is obtained against* 72·3 *cm of* *the same potentiometer wire. Find the internal resistance of the Leclanché cell.*

Let the Leclanché cell have an e.m.f. of E and an internal resistance of r. On open circuit the p.d. between the terminals (X, Y) of the cell

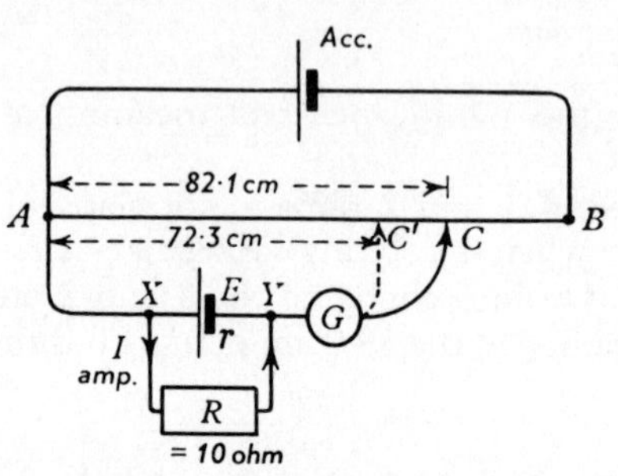

Fig. 45

is equal to the p.d. between the points A and C of the potentiometer wire (see diagram). Now since by the potentiometer principle potential drop along the wire is proportional to the length tapped off, we have

$$E = \text{p.d.}_{\text{A–C}_1} \propto 82{\cdot}1 \qquad \text{(i)}$$

On connecting a resistance R across the cell, the current I taken from it will be $\dfrac{E}{R+r}$, and the p.d. between the points X and Y is $IR = \dfrac{ER}{R+r}$ $= \dfrac{E \times 10}{10+r}$ (since $R = 10$ ohm). This p.d. is equal to the p.d. between A and C' of the potentiometer wire, hence

$$\frac{E \times 10}{10+r} = \text{p.d.}_{\text{A–C}^1} \propto 72{\cdot}3 \qquad (2)$$

Dividing equation (1) by equation (2) we get

$$\frac{10+r}{10} = \frac{82{\cdot}1}{72{\cdot}3}$$

from which

$$r = 10 \times \frac{9{\cdot}8}{72{\cdot}3} = \underline{1{\cdot}36 \text{ ohm}}$$

26 Describe how you would compare the values of two unknown resistances using a potentiometer.

A steady current is passed through two resistors, P and Q, arranged in series. With leads taken from the ends of P a balance is obtained against 22·1 cm of a given potentiometer wire. When the leads are transferred to the ends of Q, a balance against 56·0 is obtained, whereas if the leads are taken from across P and Q together, 78·6 cm of the wire is needed to secure a potentiometric balance. How do you account for these observations, and what is the value of Q if P has a resistance of 10 ohm?

27 Describe a potentiometer method for finding the internal resistance of a cell.

When measured by a calibrated potentiometer the e.m.f. of a cell is found to be 1·45 volt. The terminals of the cell are now joined through a 20-ohm resistor and the p.d. between its terminals thus connected is found to be 1·25 volt when measured with the potentiometer and 1·20 volt when measured by a voltmeter. Find the internal resistance of the cell and the resistance of the voltmeter.

28 Describe how you would use a potentiometer for the measurement of current.

An ammeter, a standard 1-ohm resistor and a variable resistor are connected in series across an accumulator, the current being adjusted so that the pointer of the ammeter is exactly over the 1-amp reading on the scale. Leads taken from the ends of the standard 1-ohm coil to a potentiometer produce a balance of potential when tapping across 47·1 cm of the wire. If a Weston cadmium cell, of e.m.f. 1·0183 volt, balances against 56·7 cm of the potentiometer wire, what is the error in the ammeter reading at the 1-amp mark?

29 Explain the action of the potentiometer and describe its use in (*a*) comparing the e.m.f.s of two cells, (*b*) measuring a small potential difference of the order of 1 mV.

A cell of e.m.f. 2 volt and internal resistance 0·5 ohm is used to send a current through a metre-long potentiometer wire which has a resistance of 3·5 ohm. Across what length of this potentiometer wire would a Daniell cell of e.m.f. 1·1 volt balance?

30 Two resistance boxes, A and B, are connected in series across a high-tension battery. Leads from the ends of the box A are taken to a potentiometer wire, a balance being obtained across 35·1 cm of this wire when the resistances in A and B are 50 and 9950 ohm respectively. If a Daniell cell, of e.m.f. 1·05 volt, balances across 51·8 cm of the potentiometer wire, calculate the e.m.f. of the high-tension battery. (You may ignore the resistance of the high-tension battery.)

31 You are provided with a metre-long potentiometer wire of resistance 3 ohm, two plug resistance boxes, A and B, each variable in steps of 1 ohm up to a maximum of 1000 ohm, and a freshly charged accumulator giving 2·1 volt. What circuit arrangement would you adopt in order to obtain a potential drop of 5 micro-volt per mm on the potentiometer wire? A Weston cadmium standard cell of e.m.f. 1·0183 volt is now to be balanced against this potentiometer. Explain how you would do this without altering the sensitivity of the potentiometer and find the possible balance points on the wire.

32 Describe how you would measure the value of a resistance of the order of a thousandth of an ohm by a potentiometer method. Give a sketch of your circuit and indicate approximate values of any necessary ancillary resistors, e.m.f.s of cells, etc., that you may require.

Thermo-electricity

33 Give a circuit diagram and full experimental detail of a method by which you could accurately ascertain the e.m.f. of a thermocouple of the order of 4 mV.

One junction of a thermo-couple of two dissimilar metals is maintained in melting ice whilst the other junction is inserted in a sand bath (with a thermometer). When the sand bath is at a temperature of 100°C, the e.m.f. between the junctions is 1·5 millivolt, and again, when the temperature is 200°C, the e.m.f. is 2·0 millivolt. What will be the thermo-electric e.m.f. between the junctions when the temperature difference is 150°C?

34 What is meant by (*a*) the *Seebeck effect,* (*b*) the *Peltier effect,* (*c*) *neutral temperature?*

The thermo-electric e.m.f. for a copper-iron couple is given by the expression

$$e\ (\text{micro-volt}) = 13{\cdot}94\, t - 0{\cdot}021\, t^2$$

where t is the temperature of the hot junction in °C, the cold junction being maintained at 0°C. Find

(i) the maximum e.m.f. for the couple,
(ii) the neutral temperature,
(iii) the temperature for which the e.m.f. is 2 millivolt.

35 Define *thermo-electric power.*

The thermo-electric power against metal A for a metal B at any given temperature t°C is given, in micro-volt per °C, by the expression $15{\cdot}50-0{\cdot}0525t$, whilst the corresponding expression for another metal C, against A, is $1{\cdot}50+0{\cdot}0075t$. What will be the e.m.f. for a thermo-couple comprised of metals B and C when the junctions are at temperatures of 10°C and 100°C?

The chemical effect of current. Cells

36 State Faraday's laws of electrolysis and describe experiments to verify them.

A Daniell cell supplies a current of 0·2 ampere for a period of 10 minutes. Find how much copper is deposited and how much zinc is consumed in this time. (Electro-chemical equivalent of copper is $0{\cdot}329\times10^{-6}$ kg C^{-1}; relative atomic masses of copper and zinc are 63·5 and 65·5 respectively.)

37 An ammeter and a silver voltameter are connected in series with an accumulator battery and a variable resistor which is adjusted until the ammeter records a current of 0·2 ampere. If the current is allowed to run for 15 minutes and 0·196 g of silver is deposited in this time, calculate the error in the ammeter reading. (Electro-chemical equivalent of silver $= 1{\cdot}118\times10^{-6}$ kg C^{-1}.)

38 Give an account of the electrical conduction of electrolytes in terms of the ionic theory and show how Faraday's laws of electrolysis may be deduced from this theory.

Calculate the number of atoms in 1 g of hydrogen given that the electro-chemical equivalent of hydrogen is $1{\cdot}044\times10^{-8}$ kg C^{-1} and the electronic charge is $1{\cdot}59\times10^{-19}$ C.

39 Using the data given below, calculate the e.m.f. of a Daniell cell assuming that the electrical energy is derived from the heat liberated during the chemical reactions taken place in it:

Energy liberated when 1 g of zinc is dissolved by sulphuric acid into solution = 6779 joule.

Energy absorbed when 1 g of copper is deposited from copper sulphate solution = 3687 joule.

Electro-chemical equivalents of zinc and copper are respectively $0{\cdot}34\times10^{-6}$ and $0{\cdot}33\times10^{-6}$ kg C^{-1}.

Worked example

40 *A heating coil and a copper voltameter are connected in series in a* *given circuit. The heating coil has a resistance of* 5 *ohm and is immersed in* 100 *g of water contained in a calorimeter of negligible heat capacity. When a current passes in the circuit a temperature rise of* 15°*C is recorded in the calorimeter in a period of* 10 *minutes. What mass of copper is deposited at the cathode of the voltameter in this time? (Electro-chemical equivalent of hydrogen* $= 1{\cdot}044 \times 10^{-8}$ *kg* C^{-1}*; relative atomic mass of copper* $= 63{\cdot}5$*; specific heat capacity of water* $= 4{\cdot}2 \times 10^3$ *J kg*$^{-1}$ *K*$^{-1}$.

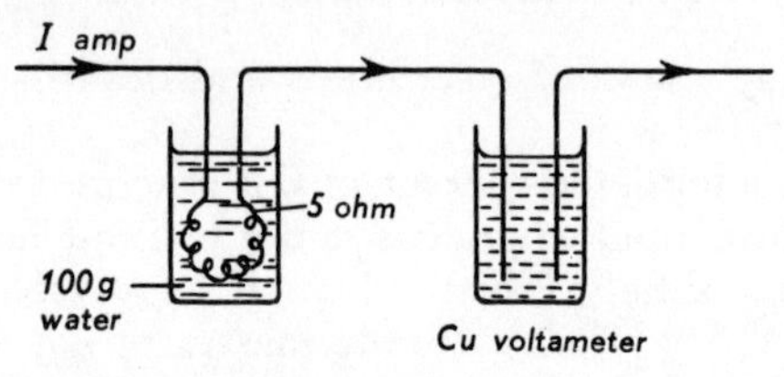

Fig. 46

Let the current passing round the circuit be I, then the heat developed in the heating coil in 10 minutes

$$= I^2Rt = I^2 \times 5 \times 600$$
$$= \text{heat gained by water (ignoring heat losses)}$$
$$= 0{\cdot}1 \times 4{\cdot}2 \times 10^3 \times 15 = 6{\cdot}3 \times 10^3 \text{ J}$$

Thus $$I = \sqrt{\frac{6{\cdot}2 \times 10^3}{5 \times 600}} = \underline{1{\cdot}45 \text{ A}}$$

The passage of this current for 600 seconds through the copper voltameter means that a quantity of electricity equal to $1{\cdot}45 \times 600$ coulombs is conveyed by the electrolyte. Hence mass of copper deposited at the cathode = E.C.E. of copper × coulombs conveyed.

Now $$\frac{\text{E.C.E. of copper}}{\text{E.C.E. of hydrogen}} = \frac{\text{chemical equivalent of copper}}{\text{chemical equivalent of hydrogen}}$$

$$= \frac{31{\cdot}75}{1} \text{ (applying Faraday's 2nd law of electrolysis)}$$

i.e. E.C.E. of copper $= 31{\cdot}75 \times 1{\cdot}044 \times 10^{-8}$

Hence mass of copper deposited

$$= 31{\cdot}75 \times 1{\cdot}044 \times 10^{-8} \times 1{\cdot}45 \times 600$$
$$= \underline{2{\cdot}89 \times 10^{-4} \text{ kg}}$$

41 Calculate the least e.m.f. needed to electrolytically decompose water given that the heat of combustion of hydrogen in oxygen is $1{\cdot}43 \times 10^8$ joules for every kilogram of hydrogen burnt, and that the

passage of one coulomb of electricity through acidulated water decomposes $9{\cdot}4 \times 10^{-8}$ kg of water.
Describe the details of some experimental method of finding this minimum e.m.f.

42 Describe some modern form of accumulator (storage battery) with a critical account of its advantages and disadvantages.

A d.c. dynamo with an output of 25 volt and having an internal resistance of 0·5 ohm is being used to charge a battery of twelve 2-volt accumulators arranged in series. If the internal resistance of each accumulator is 0·05 ohm, what is the charging current used? If it takes 22 hours to fully charge the accumulators, what is the capacity of the battery? (Assume a steady charging current throughout.)

The heating effect of a current. Electrical energy and power

43 Two wires, A and B, with lengths in the ratio 3 : 1, diameters in the ratio of 1 : 2, and with resistivities in the ratio 1 : 20 are joined in parallel across an accumulator of e.m.f. 2 volt and negligible resistance. What is the ratio of heat production in the two wires? If the total rate of heat production in the two wires is 2·1 Js^{-1}, find the resistance of each of the wires.

44 State Joule's laws of electrical heating.

Two heating coils, A and B, with resistances of 20 and 50 ohm respectively, are connected (*a*) in series, (*b*) in parallel across d.c. mains of fixed voltage. What is the ratio of the heat generated in the coils in the two cases? Find also the relative rates of heat generation in coil A in the two cases.

Worked example

45 *A 200-watt electric lamp is immersed in 1 kg of water contained in a copper calorimeter of mass 0·5 kg. If 20 per cent of the energy supplied is wasted as heat losses, find how long it will take to raise the temperature of the water from 10°C to 50°C. (Take the specific heat capacities of water and copper as $4{\cdot}185 \times 10^3$ and $0{\cdot}381 \times 10^3$ $J\ kg^{-1}\ K^{-1}$ respectively.)*

Let the required time interval = t seconds, then, since 1 watt = 1 joule per second, the 200-watt lamp will transform $200t$ joules during the period. Four-fifths of this energy is available to heat the water.

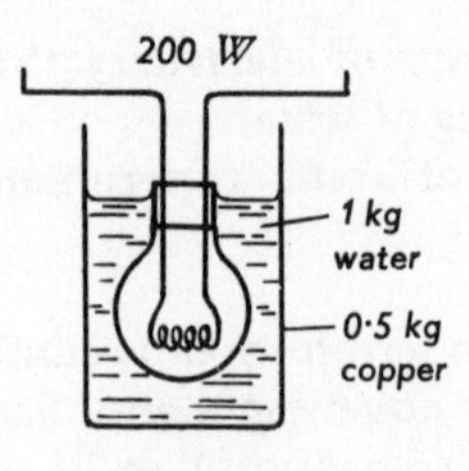

Fig. 47

Thus heat absorbed by water and calorimeter

$$= \tfrac{4}{5} \times 200t \text{ J} \qquad (1)$$

Now the total heat capacity of calorimeter and contents

$$= (1 \times 4{\cdot}185 + 0{\cdot}5 \times 0{\cdot}381)10^3 = 4{\cdot}375 \times 10^3 \text{ JK}^{-1}$$

and the heat required to produce a temperature rise from 10 to 50°C is

$$4{\cdot}375 \times 10^3 \times 40 = 175 \times 10^3 \text{ J} \qquad (2)$$

Hence from (1) and (2) we have

$$\tfrac{4}{5} \times 200t = 175 \times 10^3$$

from which

$$t = \frac{175 \times 10^3 \times 5}{4 \times 200}$$

$$= \underline{1094} \text{ seconds} = \underline{18 \text{ min } 14 \text{ sec}}$$

46 Give the details of a calorimetric method of finding the value of an electrical resistance.

A coil of wire is immersed in 200 g of water contained in a copper calorimeter of mass 100 g. When a current of 2 amperes is passed through the wire the temperature of the water is found to rise 20°C in 3 minutes. What is the resistance of the wire? If the resistance of the coil is found to be 28·0 ohm from bridge measurements, what were the percentage heat losses in the previous experiment? (Specific heat capacities of water and copper are $4{\cdot}185 \times 10^3$ and $0{\cdot}381 \times 10^3$ J kg^{-1} K^{-1} respectively.)

47 A 200-volt d.c. generator transmits power to a distant factory through cables of total length 5 km. If the cables have a resistance of 0·08 ohm per km, what is the efficiency of transmission when the cable current is 150 ampere? How would the efficiency be altered by transmitting the same power at 500 volt?

48 A steady stream of water flows along a cylindrical glass tube down the centre of which runs a length of constantan wire of resistance 2·45 ohm. The temperature of the water on entering the tube is 10·0°C and its rate of flow is 52·5 g per min. What will be the temperature of the outflowing water when the current in the wire is 2·5 amp? (Specific heat capacity of water $= 4{\cdot}185 \times 10^3$ J kg^{-1} K^{-1}. Ignore heat losses.)

Worked example

49 *Calculate the steady temperature attained by a copper wire carrying a current of 5 amp, using the following data: temperature of the surrounding air = 10°C; diameter of wire = 1 mm; emissivity of surface of wire = 9·22 J m^{-2} K^{-1} s^{-1}; resistivity of copper $= 1{\cdot}8 \times 10^{-8}$ ohm-metre units; temperature coefficient of resistance of copper = 0·0043 K^{-1}.*

Let the temperature attained by the wire be θ°C. Then, if R_{10} is the resistance of the wire at 10°C, its resistance at θ°C will be $R_{10}[1+0{\cdot}0043(\theta-10)]$. Thus the heat generated per second with a current of 5 amp will be $5^2 \times R_{10}\{1+0{\cdot}0043(\theta-10)\}$

$$= \frac{5^2 \times 4 \times 1{\cdot}8 \times 10^{-8} l}{\pi \times (0{\cdot}001)^2} \times \{1+0{\cdot}0043(\theta-10)\} \qquad (1)$$

where l is the length of the wire.

At the steady temperature the heat generated is lost by radiation and convection from the surface of the wire, and, assuming Newton's law of cooling, the heat thus lost per second

$$= \varepsilon\, s\,(\theta-10)$$

where ε is the emissivity and S the surface area of the wire. Thus the heat lost per second from the surface of the wire

$$= 9{\cdot}22 \times \pi \times 0{\cdot}001 \times l\,(\theta-10) \qquad (2)$$

Hence equating (1) and (2) we get

$$1+0{\cdot}0043(\theta-10) = \frac{\pi^2 \times (0{\cdot}001)^3 \times 9{\cdot}22\ (\theta-10)}{25 \times 4 \times 1{\cdot}8 \times 10^{-8}}$$

$$= 0{\cdot}0506(\theta-10)$$

from which $$\theta-10 = \frac{1}{0{\cdot}0463} = 21{\cdot}6$$

giving $$\theta = \underline{31{\cdot}6°\text{C}}.$$

50 Discuss the factors which determine the steady temperature attained by a current-bearing wire.

A length of copper wire, carrying a current of 2 ampere, attains a steady temperature of 25°C, the room temperature being 10°C. What will be the temperature attained by the wire when the current is raised to 3 ampere? The temperature coefficient of copper = 0·0042 K^{-1}. (Ignore the expansion of the wire.)

51 Assuming that the heat losses from a current-bearing wire may be evaluated by means of Newton's law of cooling for moderate temperature rises, and ignoring changes of resistance with temperature, obtain an expression for the steady surface temperature attained by such a wire. Hence show, for low melting-point fuse wires, that $I \propto d^{\frac{3}{2}}$, where I is the fusing current and d the diameter of the wire.

Calculate the approximate value of the fusing current of pure tin wire of diameter 0·7 mm from the following data. Melting point of tin = 230°C, room temperature = 10°C, resistivity of tin = 15×10^{-8} Ω m emissivity of surface of wire = 9·22 J m^{-2} K^{-1} s^{-1}.

52 A short-circuit current of 4000 ampere suddenly passes through the coils of an electromagnet wound with copper wire of diameter 3 mm. How long does the current flow before the windings begin to melt? State any assumptions made in your calculation. (Melting point of copper = 1080°C; room temperature = 10°C; density of copper $= 8{\cdot}9 \times 10^3$ kg m^{-3}; specific heat capacity of copper = $0{\cdot}4 \times 10^3$ J kg^{-1} K^{-1}; resistivity of copper = $1{\cdot}8 \times 10^{-8}$ Ω m.)

53 Ignoring thermal expansion effects, and assuming the resistance of a tungsten filament lamp varies directly as the absolute temperature, and that the energy radiated from its surface conforms to Stefan's fourth power law, show that the relation between the current (I) passing through the filament and the potential difference (V) applied across it is given by the expression

$$I = kV^{3/5} \text{ where } k = \text{constant.}$$

Give the details of an experiment you would perform to verify this relationship.

Electro-magnetism. Galvanometers

Some formulae and units	*units*
Magnetic space constant	
or permeability of free space = $\mu_0 = 4\pi \times 10^{-7}$	H m^{-1}

Relative permeability $= \mu_r = 1$ for vacuum (or air)
Absolute permeability $= \mu = \mu_r \mu_0$ — $\mathrm{H\,m^{-1}}$

Magnetic flux $= \phi$ — Wb

Magnetic flux density $= B = \dfrac{\phi}{A}$ — $\mathrm{Wb\,m^{-2}}$ or T

Force on current-bearing conductor in magnetic field

$= F = BIl$ (or, general, $= BIl \sin\theta$) — N

Couple on current-bearing coil in magnetic field
$= T = NBAI \cos\theta$ — N-m

Magnetic moment of a coil $= m = NAI$ — $\mathrm{A\,m^2}$

Flux density due to current-bearing element

(Biot-Savart formula) $= dB = \dfrac{\mu_0}{4\pi}\dfrac{Idl \sin\theta}{r^2}$ — $\mathrm{Wb\,m^{-2}}$

Flux density at distance d from 'long' current-bearing wire

$= B = \dfrac{\mu_0 I}{2\pi d}$ — $\mathrm{Wb\,m^{-2}}$

Force between two current-bearing wires parallel wires
(on length l of either)

$= F = \dfrac{\mu_0 I_1 I_2 l}{2\pi d}$ — N

Flux density at centre of current-bearing coil

$= B = \dfrac{\mu_0 NI}{2r}$ — $\mathrm{Wb\,m^{-2}}$

Flux density inside a 'long' solenoid

$= B = \dfrac{\mu NI}{l}$ — $\mathrm{Wb\,m^{-2}}$

Flux density on axis of a current-bearing coil

$= B = \dfrac{\mu_0 N r^2 I}{2(r^2+x^2)^{3/2}}$ — $\mathrm{Wb\,m^{-2}}$

Flux in magnetic circuit

$= \phi = \dfrac{\text{magneto-motive force}}{\text{reluctance}}$

$= \dfrac{NI}{l/\mu_0 \mu_r A}$ — Wb

Pull between poles of a magnet (per unit of pole area)

$$= \frac{1}{2}\frac{B^2}{\mu_0} \qquad \text{N m}^{-2}$$

Self inductance of a coil

$$= L = \frac{N\phi}{I} = \frac{N^2\mu A}{l} \qquad \text{H}$$

Self induced e.m.f. $= e = -L\frac{dI}{dt}$ V

Energy associated with an inductive coil

$$= \tfrac{1}{2}LI^2 \qquad \text{J}$$

54 Obtain an expression for the magnetic field intensity at a point a given distance away from a long current-bearing conductor.

Use your expression to find the current in an overhead conductor if it produces a field intensity equal to that of the Earth's horizontal component ($0{\cdot}18 \times 10^{-4}$ Wb m^{-2}) at ground level 10 m below the cable (running parallel to the ground). (Permeability of free space = $4\pi \times 10^{-7}$ H m^{-1}.)
Worked example

55 *A current is passing down a long vertical wire situated in the Earth's magnetic field. If a neutral point is located at a perpendicular distance of 10 cm from the wire, what is the value of the current? ($B_0 = 1{\cdot}8 \times 10^{-5}$ Wb m^{-2} and $\mu_0 = 4\pi \times 10^{-7}$ H m^{-1}.)*

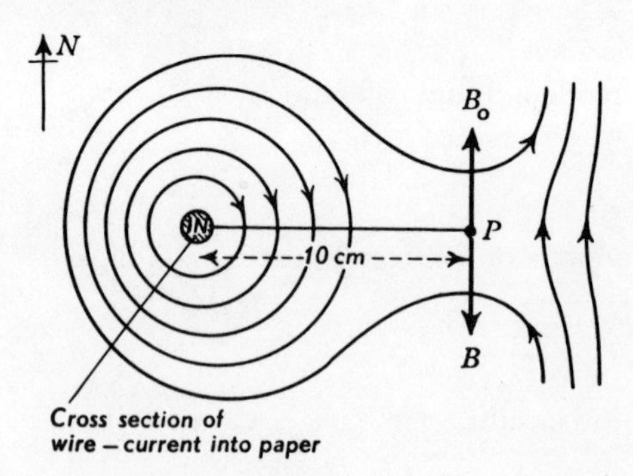

Fig. 48

The diagram shows a horizontal section of the wire and the configuration of the flux flow due to the current and the Earth's field. A point of zero flux intensity (neutral point) is shown at P, 10 cm due east of the wire at which position the flux intensities due to the wire (B) and the Earth's horizontal field (B_0) are equal and opposite. Thus,

$$B = B_0 = 1{\cdot}8 \times 10^{-5}\ \mathrm{Wb\ m^{-2}}$$

Now the flux intensity B at a distance d from a long wire carrying a current $I = \dfrac{\mu_0 I}{2\pi d}$. Hence, in this case,

$$B = \frac{4\pi \times 10^{-7} \times I}{2\pi \times 0{\cdot}1} = 2I \times 10^{-6}$$

Thus we have

$$2I \times 10^{-6} = 1{\cdot}8 \times 10^{-5}$$

giving

$$I = \underline{9}\ \text{ampere}$$

56 Give an expression for the force per metre length on a straight conductor carrying a current I when placed in a transverse uniform magnetic field of flux density B. Indicate on a diagram the special relationship between current, magnetic field and the force acting.

Two parallel conducting cables each carry currents (in the same direction) of 50 A. If they are situated 2 m apart, calculate the force between them per metre run of either cable. ($\mu_0 = 4\pi \times 10^{-7}\ \mathrm{H\ m^{-1}}$.)

57 An overhead cable, running in an E–W direction, carries a current of 100 A d.c. What is the force per metre length of this cable produced on it by the earth's magnetic field if, at the position in question, the angle of dip is 60° and the horizontal component of the Earth's field is $0{\cdot}18 \times 10^{-4}\ \mathrm{Wb\ m^{-2}}$?

58 The interpole field of a large electromagnet has a flux density of $2{\cdot}0\ \mathrm{Wb\ m^{-2}}$. A 5-cm length of wire carrying a current of 15 mA lies in this field. Calculate the force on the wire when it is (*a*) perpendicular to the flux, (*b*) parallel with the flux, (*c*) makes an angle of 30° with the flux. ($\mu_0 = 4\pi \times 10^{-7}\ \mathrm{H\ m^{-1}}$.)

59 Two long straight wires A (carrying 2 ampere) and B (carrying 3 ampere—in the same direction as the current in A) run in parallel directions 10 cm apart. Considering a cross-section of the arrangement of the wires, find the position at which the field intensity due to the current-bearing conductors is zero. Where will this position be relative to the wires if B's current is reversed? (Neglect the effect of the Earth's field.)

60 A long straight wire is tightly stretched between two terminal blocks so that it lies along the magnetic meridian and parallel to the bench surface. A deflection magnetometer placed with its needle 10 cm vertically below the mid point of the wire records a deflection of 30° when a current of 5·2 A is passed along the wire. From these observations obtain a value for the Earth's horizontal magnetic field. ($\mu_0 = 4\pi \times 10^{-7}$ H m^{-1}.)

61 A very long wire A carrying a current of 5 ampere lies parallel with, and distant 2 cm from, a shorter wire B of length 10 cm and carrying a current of 2 ampere in the same direction as A's current. Calculate the force exerted by A on B. Find also the work done in separating B a further 2 cm from A.

62 Two square-shaped single wire coils each carry a current of 2 A circulating in the same direction through each coil. Calculate the force required to hold the two coils at a distance of 0·4 cm apart if the length of the side of the square of each coil is 5 cm.

63 The 'force element' of a current balance consists of a 4-cm long horizontal copper wire piece which is supported at the end of a lever arm 20 cm long through which the current passes to the element. The balance is set up so that the element is positioned between the pole pieces of a large electromagnet and it is found that, when a current of 2 A is passed through the element a 2 gramme rider mass has to be positioned 15 cm from the pivot along the counterpoising arm of the balance. Calculate a value for the field intensity between the pole pieces of the magnet. (Take g as 9·80 m s^{-2}.)

64 The 'speech coil' of a small permanent magnet loudspeaker unit has 200 turns of mean diameter 2 cm and is arranged so that it can move in an annular spacing in the speaker magnet. In this way the field of the magnet is always radial to the coil. If the strength of this field is 0·2 Wb m^{-2}, calculate the force on the coil when a current of 5 mA passes through it.

65 Obtain an expression for the magnetic flux density at the centre of a current-bearing coil.

A circular coil having 20 turns of mean radius 10 cm is capable of rotation about a vertical diametral axis. Describe how you would position the coil and determine the current through it if it required to neutralize the Earth's horizontal component (of strength $0{\cdot}18 \times 10^{-4}$ Wb m^{-2}) in the plane of the coil.

66 A flat circular coil of 20 turns of mean radius 10 cm is mounted with the plane of its face vertical and with its axis running along the E–W line. At a point 20 cm along the axis from the centre of the coil is positioned the needle of a deflection magnetometer. Find what current passing round the coil will yield a deflection of 45° of the megnetometer needle. (Earth's horizontal component $= 0{\cdot}18 \times 10^{-4}$ Wb m^{-2}.)

67 Two circular current-bearing coils are mounted symmetrically one inside the other and in such a way that the inner one can be rotated about a vertical diametral axis. The outer coil has 50 turns of mean radius 7·5 cm whilst the inner coil has 100 turns of mean radius 5 cm. The coils are wired in series and carry a current of 2 A. What is the magnitude and direction of the resultant magnetic field at the common centre of the coils (*a*) when their magnetic axes are coincident, (*b*) when the inner coil is reversed from its position in (*a*), (*c*) when the axis of the inner coil makes 30° with its original position? Ignore effects due to the Earth's field.

68 Derive an expression for the magnetic field along the axis of a current-bearing coil.

It is required to produce a region of uniform magnetic flux by using two circular current-bearing coils. Explain how you would arrange the coils and calculate the flux density in the uniform region if each coil has 50 turns of mean radius 7·5 cm and a current of 2 A passes through each.

69 Two similar coils, each of 20 turns of wire of radius 10 cm are arranged co-axially with their planes in the meridian and with their centres 10 cm apart. The coils are wound in series and when a current passes through them a magnetic needle, suspended mid-way between the coils, is deflected through an angle of 30°. What is the strength of the current? ($B_0 = 1{\cdot}8 \times 10^{-5}$ Wb m^{-2}.)

70 Derive an expression for the magnetic flux density inside a long current-bearing solenoid.

Such a 'long' solenoid has a diameter of 4 cm and a length of 20 cm. It has 200 turns through which a current of 2 A is passed, and when the core space is filled with a ferromagnetic material a flux density of

0·8 Wb m^{-2} is produced in the specimen. Calculate a value for the relative permeability of the specimen.

71 Describe the essential features of a moving coil galvanometer. How is a linear relationship of current and deflection obtained with such a galvanometer?

The 100 turns of such a galvanometer are wound on a rectangular frame $2 \times 1{\cdot}5$ cm, the longer side being vertical. If the inter-pole field is 0·25 Wb m^{-2}, calculate the torque acting on the coil when a current of 5 mA passes through it.

72 A rectangular wire frame with 10 turns each of dimensions 2×3 cm is suspended from a phosphor-bronze strip in such a way that the coil can rotate about a symmetrical vertical axis in a uniform magnetic field of strength 0·5 Wb m^{-2}. Find the force on the vertical (3 cm) members of the frame when a current of 0·1 A passes round it and the torsional constant of the phosphor bronze strip of the coil sets itself with the plane of its force at 60° with the direction of the field.

73 Describe the construction and mode of action of some form of ammeter employing the magnetic effect of a current.

A given galvanometer has a resistance of 5 ohm and gives a full-scale deflection when taking a current of 15 mA. How would you adapt this instrument for use (*a*) as an ammeter with range 0–5 A, (*b*) as a voltmeter with range 0 to 10 V?

74 How would you find the current sensitivity of a moving-coil galvanometer?

Such a galvanometer has a rectangular coil consisting of 20 turns with dimensions 3×2 cm. It is suspended (with the longer side vertical) in a radial magnetic field of 0·4 Wb m^{-2} by means of a phosphor-bronze fibre which provides a restoring couple of $1{\cdot}8 \times 10^{-7}$ N m per radian of twist. What is the deflection of the coil when a current of 15 μA passes round it?

75 Describe how a ballistic galvanometer can be used to compare the capacitances of two capacitors.

A ballistic galvanometer has a sensitivity of 3·5 divisions per microcoulomb. When a capacitor is charged by momentarily connecting it across a 6-volt battery and then immediately discharged through the ballistic galvanometer, a deflection of 44·1 divisions is obtained. What is the capacitance of the capacitor?

76 Give the theory of the moving-coil ballistic galvanometer.

A moving-coil ballistic galvanometer has a quantity sensitivty of 5 divisions per micro-coulomb and the time for one complete oscillation of the suspended system when on open circuit is 5·5 second. What steady deflection would be obtained with this galvanometer when a current of 5 μA passes through the galvanometer?

77 Discuss the essential features of a ballistic galvanometer.

A 2 μF capacitor is charged by connecting it across an accumulator battery. It is then disconnected and discharged through a ballistic galvanometer which records a deflection of 33·6 divisions. The capacitor is now recharged (using the same battery), but now, on being disconnected from the battery, a high resistance is joined across the capacitor terminals for a period of 12 seconds. On removing this resistance and discharging the capacitor through the ballistic galvanometer a deflection of 4·2 divisions is obtained. What is the value of the high resistance? Develop any formulae used.

78 What do you understand by *logarithmic decrement,* and on what factors does the shrinkage of the swings of a ballistic galvanometer depend?

On being deflected a ballistic galvanometer records a first throw of 30 divisions and subsequently an eleventh throw (in the same direction as the first) of 12 divisions. Calculate a value for the logarithmic decrement of the galvanometer when working under these conditions, and obtain a value for the undamped first throw.

79 What do you understand by the term *magneto-motive force*? How is it related to the flux flow in a magnetic circuit?

An iron ring with a mean circumference length of 50 cm, and an area of cross-section of 5 cm^2, is wrapped with a toroidal coil of enamelled copper wire through which a current of 5 A is passed. If the relative permeability of the iron is 2000, and the coil has a total of 500 turns, calculate the flux produced in the iron ring.

What would be the effect on the flux of cutting an air section of width 2 mm in the specimen?

Electromagnetic induction. Inductance. Motors and dynamos

80 State the laws of electromagnetic induction and describe how they may be demonstrated by laboratory experiments.

A vertical car radio aerial is 1 m long. Calculate the steady voltage induced across it when the car is travelling at 80 km hr^{-1} in an E-W direction. (Horizontal component of the Earth's field $= 0{\cdot}18 \times 10^{-4}$ Wb m^{-2}.)

81 A metal disc of radius 10 cm spins about its axis at the rate of 15 revolutions per second. If the plane of the disc is perpendicular to a uniform magnetic field of 0·5 Wb m^{-2} which completely embraces the area of the disc, find the steady e.m.f. established between the axis and rim of the disc. Give the theory of your method.

Worked example

82 *An engine travels north at a uniform speed of 96 km hr^{-1} in a straight horizontal path. Calculate the e.m.f. set up at the ends of a conducting axle* 150 *cm long if $B_0 = 2 \times 10^{-5}$ Wb m^{-2} and the angle of dip is* 68°.

$$96 \text{ km hr}^{-1} = \frac{96 \times 1000}{60 \times 60} = \frac{80}{3} \text{ m s}^{-1}$$

Hence, in one second, the axle of the train sweeps out an area of

$$1{\cdot}5 \times \frac{80}{3} = 40 \text{ m}^2$$

The axle also cuts through the vertical component of the Earth's magnetic field which has a value of 2×10^{-5} tan 68 $= 4{\cdot}95 \times 10^{-5}$ Wb m^{-2}. Hence flux cut by axle in one second

$$= 40 \times 4{\cdot}95 \times 10^{-4}$$

$$= 1{\cdot}98 \times 10^{-3}$$

Now the induced e.m.f. = rate of cutting of the magnetic flux, thus, induced e.m.f. across the axle

$$= 1{\cdot}98 \times 10^{-3} = \underline{0{\cdot}00198} \text{ volt}$$

83 An aircraft flies at a constant height above the Earth's surface at a speed of 500 km hr^{-1}. Assuming the vertical component of the Earth's field has a value of $0{\cdot}32 \times 10^{-4}$ Wb m^{-2}, calculate the p.d. established between the wing tips of the plane if the wing span is 30 m.

84 A flat circular coil of wire with 100 turns of radius 12 cm is rotated at a constant rate of 200 revolutions per minute about a vertical diametral axis when placed in a uniform magnetic field of 5×10^{-3} Wb m^{-2} acting in a direction perpendicular to the axis of rotation of the coil. Calculate (*a*) the maximum value of the e.m.f. induced in the coil, (*b*) the instantaneous value of the induced e.m.f. when the plane of the coil makes an angle of 60° with the field direction.

85 Obtain an expression for the quantity of electricity flowing through a circuit as a result of a change of magnetic flux across it.

A flat circular coil consisting of 100 turns of fine wire of mean radius 2·0 cm and with a resistance of 25 ohm, is connected to a ballistic galvanometer having a sensitivity of 1·5 divisions per micro-coulomb and a resistance of 45 ohm. The coil is placed between the pole pieces of an electromagnet so as to be perpendicular to the field and on removing the coil a deflection of 275 divisions is registered by the ballistic galvanometer. What is the strength of the magnetic field between the pole pieces of the magnet?

86 A ballistic galvanometer is connected to an earth inductor which is set vertically with its plane perpendicular to the meridian, and on rotating the coil through 180° a deflection of 9 divisions is registered by the ballistic galvanometer. The earth inductor is now laid horizontally and on rotating it through 180° about a horizontal axis the ballistic galvanometer registers a deflection of 22 divisions. Obtain a value for the angle of dip.

87 A small search coil of 50 turns of fine wire of mean radius 1 cm is placed inside a current-bearing solenoid so that the axis of the coil is parallel with that of the solenoid. The ends of the search coil are connected to a calibrated ballistic galvanometer, the total resistance of the galvanometer circuit being 250 ohm. If, when the solenoid current is stopped, a discharge of 2μC is recorded by the galvanometer, find the value of the field inside the solenoid.

88 A small search coil is placed between the pole pieces of an electro-magnet and it is found that, on sharply withdrawing the coil from the field, a ballistic galvanometer connected in circuit with the search coil records a deflection of 400 divisions. The same search coil is now placed inside a 'long' solenoid wound with 20 turns per cm and carrying a current of 3 A. On suddenly switching off the current in the solenoid, a deflection of 30 divisions is now recorded on the galvanometer scale. Calculate the interpole field strength of the electro-magnet. (You may assume that the search coil is positioned with its face perpendicular to the flux in the two cases and that the electrical circuit conditions of the ballistic galvanometer circuit remain unchanged throughout the experiment.)

89 What do you understand by the term *self-inductance*? Define the unit in which it is measured.

A solenoidal coil has 200 turns wound on a former of radius 2 cm. If the windings occupy a total length of 15 cm of the former, calculate a value for the self inductance of the coil.

What would be the total flux produced in this coil when a current of 3 A is passed through its windings?

90 Describe the construction and explain the action of an electrical transformer.

A transformer is required to step up from 220 volt to 33 000 volt. If the primary winding consists of 132 turns, how many turns are there in the secondary? Find also the current in the secondary coil when a current of 3·0 ampere passes through the primary.

Worked example

91 *A 'long' solenoid, wound with* 20 *turns per cm run, has a radius of* 3 *cm and passes a current of* 5 *A. Round the middle of this solenoid is wound a secondary coil having a total of* 100 *turns. What is the p.d. established across this coil when the current in the primary solenoid is quenched in* 0·002 *second?*

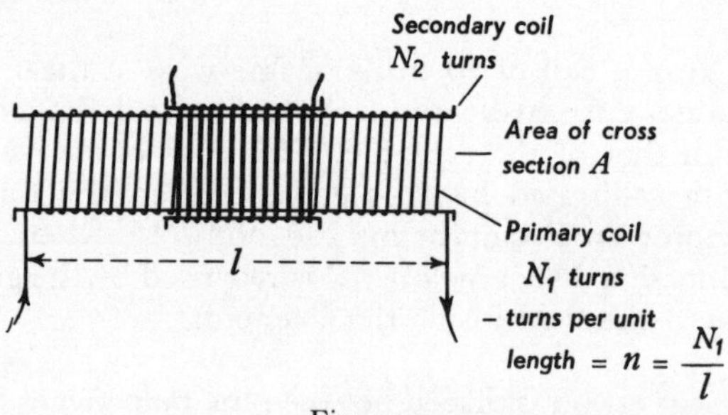

Fig. 49

The arrangement of the coils acts as a mutual conductor, the mutual inductance (M) of the secondary coil being equal to the flux linkages across it due to the passage of unit current in the primary. Now the flux density B in the primary coil due to a current I passing round its windings is $\mu_0 nI$ and the flux ϕ $(=BA)$ $=\mu_0 nIA$. For unit current this is $\mu_0 nA$ and hence the flux linkage with the secondary $= N_2\phi$ $= \mu_0 nN_2A$. Thus,

$$M = \mu_0 nN_2A = 4\pi\times10^{-7}\times\frac{20}{0{\cdot}01}\times100\times\pi(0{\cdot}03)^2$$

Now the e.m.f. (e) established across the secondary due to a rate of change of current $\left(\frac{dI}{dt}\right)$ in the primary is given by the expression

$e = -M\frac{dI}{dt}$. In this case $\frac{dI}{dt} = \frac{5}{0{\cdot}002}$ and hence

$$e = -4\pi \times 10^{-7} \times \frac{20}{0{\cdot}01} \times 100 \times \pi(0{\cdot}03)^2 \times \frac{5}{0{\cdot}002}$$

$\underline{1{\cdot}78 \text{ V}}$

92 A solenoidal coil, of 200 turns, is connected in circuit with a battery, reversing key, contral rheostat and ammeter, the rheostat being adjusted to give a current of 2 A through the coil. Round the middle of the solenoid is wrapped a separate short coil of 20 turns which is connected through a resistance to a calibrated ballistic galvanometer. When the current in the solenoidal coil is quickly reversed, the throw of the ballistic galvanometer indicates that a quantity of 5 μC of charge has passed through it. If the total resistance of the galvanometer circuit is 500 ohm, find a value for the self inductance of the solenoidal coil.

93 The primary coil of a transformer is supplied with a current of 2 A r.m.s. alternating at 50 Hz. If the transformer has a mutual inductance of 5 H, find the e.m.f. established across the terminals of the secondary coil.

94 Obtain an expression for the growth of a current in an inductive circuit.

An electro-magnet with an inductance of 5 H and a resistance of 25 ohm is connected across 100-volt d.c. mains. Calculate the time elapsing, after making the circuit, for the current to build up to 2 ampere. If the circuit can be broken in $\frac{1}{200}$ second, what is the magnitude of the self-induced e.m.f. across the terminals when the circuit is broken after the current has been established?

Worked example

95 *An inductive coil of resistance* 10 *ohm is connected across a battery of* 20 *volt and negligible resistance and on making the circuit the current is found to reach a value of* 1 *ampere after the lapse of* 0·5 *second. From this data obtain a value for the self inductance of the coil and the time elapsing to the stage where the opposing self-induced e.m.f. is* 5 *volt.*

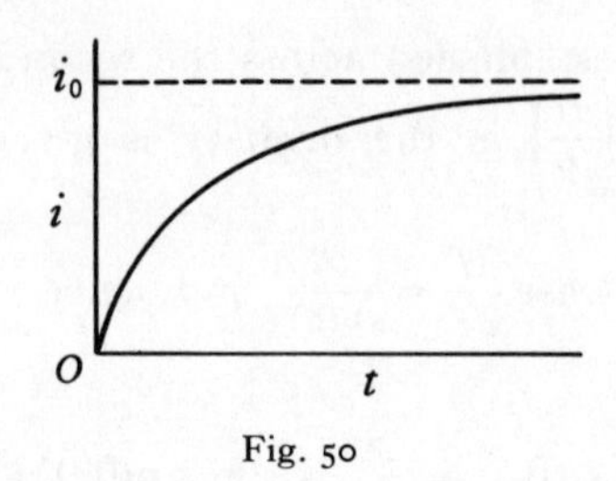

Fig. 50

The growth of the current in the coil is given by the expression

$$i = i_0(1 - e^{-\frac{R}{L}t})$$

R, L being the resistance and self inductance of the coil, and i_0 the ultimate current value in the circuit (see Fig. 50). This, clearly, is $\frac{20}{10} = 2$ A so, inserting values in the above expression we have,

$$1 = 2(1 - e^{-\frac{10}{L} \times 0 \cdot 5})$$

which reduces to $e^{5/L} = 2$

giving $L = \frac{5}{\log_e 2} = \underline{7 \cdot 21 \text{ H}}$

When the self induced e.m.f. is e, the circuit current is $i = \frac{V - e}{R}$.

Hence when $e = 5$ volt, $i = \frac{20 - 5}{10} = \left(\frac{15}{10}\right) = 1 \cdot 5$ ampere.

Accordingly, from $i = i_0(1 - e^{-\frac{R}{L}t})$ we have

$$1 \cdot 5 = 2(1 - e^{-\frac{10}{7 \cdot 2}t})$$

from which $t = \frac{\log_e 8}{10/7 \cdot 21} = \underline{1 \cdot 50 \text{ second}}$

96 Explain the effect of inductance on the growth of a current in a circuit and obtain an expression for the growth rate in terms of the quantities involved.

A self inductive coil of resistance 2 ohm and having an inductance of 5 H is connected in circuit with a 10-volt battery of negligible resistance and a depression switch. Find (*a*) the maximum current, (*b*) the greatest rate of change of current in the circuit after the switch is depressed to close the circuit.

97 A solenoidal coil is connected in series with a battery, an accumulator and a switch controlled by a milli-second timing device. On closing the circuit the current has a value of 1 ampere after an interval of 2 millisecond, but on filling the core of the coil with a ferromagnetic material, the current takes 2 second to reach a value of 1 ampere. Estimate a value for the relative permeability of the core material from these observations and comment on the value so obtained.

98 A coil with an inductance of 2 H and a resistance of 5 ohm is connected across a 20-volt battery of negligible resistance. Calculate the rate of change of current when the current is 2 ampere and the energy associated with the circuit when the current is steady. Establish any formulae used.

99 What is *mutual inductance*?

A solenoid is 30 cm long, has 800 turns, and its cross-sectional area is 10 cm^2. Closely wound on its central portion is another coil having 500 turns. Find the mutual inductance of the coils and the voltage induced in the outer coil due to a rate of change of current of 25 ampere per second in the inner coil.

Worked example

100 *A shunt-wound motor has an armature resistance of* 0·5 *ohm and field coils of resistance* 50 *ohm. When working at the rate of* 3 *H.P. on* 200-*volt d.c. mains the armature speed is* 400 *rev. per min. If an increase of load puts up the working rate to* 6 *H.P., what is the new armature speed?* (1 *H.P.* = 746 *watt.*)

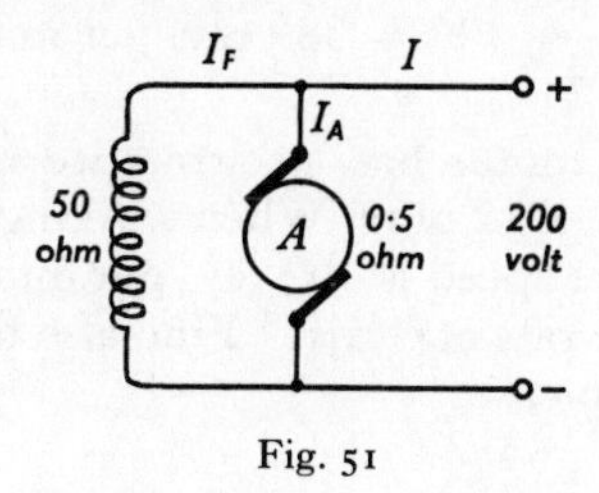

Fig. 51

Power supplied to motor = 3 × 746 watt = 200 × I where I is the current drawn from the mains.

Hence $$I = \frac{3 \times 746}{200} = 11{\cdot}19 \text{ A}$$

Now I_F (current through field coils) $= \frac{200}{5} = 4$ A

$\therefore I_A$ (current through armature) $= 11{\cdot}19 - 4 = 7{\cdot}19$ A

If now the back e.m.f. in the armature windings is e then

$$I_A = \frac{200-e}{0\cdot5} = 7\cdot19$$

from which $e = 200-3\cdot6 = \underline{196\cdot4 \text{ V}}$

When the power supplied is 6 H.P., the mains current is

$$\frac{6\times746}{200} = 22\cdot38 \text{ A.}$$

The field current is still 4 A, and hence the new armature current

$$= 22\cdot38-4 = 18\cdot38 \text{ A}$$

If the new back e.m.f. is e′ we have

$$18\cdot38 = \frac{200-e'}{0\cdot5}$$

from which $e' = 200-9\cdot19 = \underline{190\cdot8 \text{ V}}$

Now the back e.m.f $\propto$ rate of cutting of magnetic flux
$\propto$ speed $\times$ magnetic flux intensity
$\propto$ speed $\times$ field current
$\propto$ speed (since field current is constant)

Hence $$\frac{196\cdot4}{190\cdot8} = \frac{\text{old speed}}{\text{new speed}}$$

or $$\text{new speed} = \frac{190\cdot8}{196\cdot4}\times400 = \underline{389} \text{ rev. per min.}$$

101 A series-wound motor has an armature resistance of 0·5 ohm and field coils of resistance 2 ohm. When working at 3 H.P. on 200-volt d.c. mains the armature speed is 400 rev. per min., what is it when the motor is working at the rate of 6 H.P.? Find also the electrical efficiency of the motor in the two cases.

102 Discuss the relative merits of shunt-wound and series-wound d.c. motors. Why is it necessary to include a resistance in series with large shunt-wound motors? Give a sketch and brief description of some form of motor starter.

103 A shunt-wound motor has an armature resistance of 0·5 ohm and operates on 200-volt d.c. mains. Under a certain load the armature current is 4 ampere and the motor then runs at a speed of 400 rev. per min. When under full load the motor speed falls to 390 rev. per min. What is the increase in current drawn from the mains?

Alternating current circuits

Some fundamental properties of a.c. circuits

Peak value of alternating e.m.f. $= E_0$

r.m.s. value of alternating e.m.f. $= E_{\text{r.m.s.}}$

For sinusoidal a.c. $E_{\text{r.m.s.}} = \dfrac{E_0}{\sqrt{2}}$ V

Similarly for currenr $I_{\text{r.m.s.}} = \dfrac{I_0}{\sqrt{2}}$ A

Pulsatance of a.c. supply $= \omega$ rad s^{-1}

Frequency of a.c. supply $= f = \dfrac{\omega}{2\pi}$ Hz or s^{-1}

Reactance of a capacitative component $= X_C = \dfrac{1}{C\omega}$

$$= \frac{1}{2\pi fC} \quad \Omega$$

Reactance of an inductive component $= X_L = L\omega$

$$= 2\pi fL \quad \Omega$$

Impedance of capacitance and resistance in series

$$= Z = \sqrt{R^2 + X^2{}_C} \quad \Omega$$

Phase angle of such a circuit (current leading)

$$= \phi = \tan^{-1}\frac{X_C}{R}$$

Impedance of a resistance and inductance in series

$$= Z = \sqrt{R^2 + X_L^2} \quad \Omega$$

Phase angle of such a circuit (current lagging)

$$= \phi = \tan^{-1}\frac{X_L}{R}$$

General a.c. series with resistance, capacitance and inductance

$$\text{– Impedance} = Z = \sqrt{R^2 + (X_L \sim X_C)^2} \quad \Omega$$

$$\text{– phase angle of such a circuit} = \phi = \tan^{-1}\frac{X_L \sim X_C}{R}$$

Power factor of an a.c. circuit $= p.f. = \cos\phi$

Resonant frequency of an a.c. circuit

$$= f_r = \frac{1}{2\pi\sqrt{LC}} \quad \text{Hz or } s^{-1}$$

104 Establish a relation between the *peak value* and *root mean square value* of an alternating current and define the terms involved.

A heating coil is immersed in a liquid contained in a beaker. Find what the r.m.s. value of an a.c. current must be to give an initial rate of rise of temperature of the liquid which is 5 times that produced by a current of 3 amp d.c. when passed through the resistor under the same conditions.

105 A choke coil of negligible resistance takes a current of 5 ampere when connected to 200-volt, 50-cycle mains. What current would it take if connected to 150-volt, 40-cycle mains?

106 Define the terms *reactance* and *impedance* as used in a.c. circuits.

A coil has an inductance of 0·2 H. What is its reactance when used on 50-cycle supply mains? If the voltage of the mains is 230, what current would be taken when the above coil is connected in series with a resistance of 50 ohm across the mains?

107 Describe an ammeter suitable for the measurement of a.c. current.

An electric heater of resistance 25 ohm is supplied from 230-volt, 50-cycle alternating current mains. Determine: (*a*) the average power, (*b*) the maximum instantaneous power.

108 Define the term *reactance* in an a.c. circuit and indicate the different types of reactance in such a circuit with their effect on the current.

A source of 100 r.m.s. volts alternating at 50 Hz is applied across a 5 μF capacitor and a non-inductive resistor of 200 ohm arranged in series. Find:

(*a*) the r.m.s. value of the circuit current,

(*b*) the p.d. across the capacitor,

(*c*) the rate of energy supply from the source.

109 When placed across an a.c. supply of frequency 500 Hz, an inductive coil, of resistance 10 ohm, passes a current of 4 mA r.m.s. The frequency of the supply is now increased, at constant output voltage, to 1000 Hz and the current now passed by the coil falls to 3 mA r.m.s. From these observations find the coil inductance.

110 In the above problem find the capacitance of a capacitor connected in series with the inductive coil which will draw a maximum current from the mains at 1000 Hz. What is the value of this current when the voltage is 50 mV r.m.s.?

111 What do you understand by the term 'power factor' in an a.c. circuit. How is it determined and what is its significance in electrical power transmission?

A reactive coil comprising a resistive component R and an inductive component L is connected across an a.c. mains supply of 200 volt r.m.s. at 50 Hz. The current is observed to be 0·25 ampere r.m.s. with a phase lag of 60° behind the supply volts.

Obtain the R and L values of the coil.

In order to bring the current fully in phase with the supply volts a capacitor C is connected in parallel with the coil. What capacitance must C have and what is the current now taken from the mains supply?

Worked example

112 *A heating coil, of resistance* 100 *ohm, a choke (of negligible resistance) with an inductance of* 0·5 *Henry and a capacitor of capacitance* 15 *micro-farad, are connected in series across* 200*-volt,* 50*-cycle mains. Calculate (a) the impedance of the circuit, (b) the current taken from the mains, (c) the potential drop across the various components in the circuit, (d) the power factor of the circuit and (e) the energy converted by the heating coil in one cycle.*

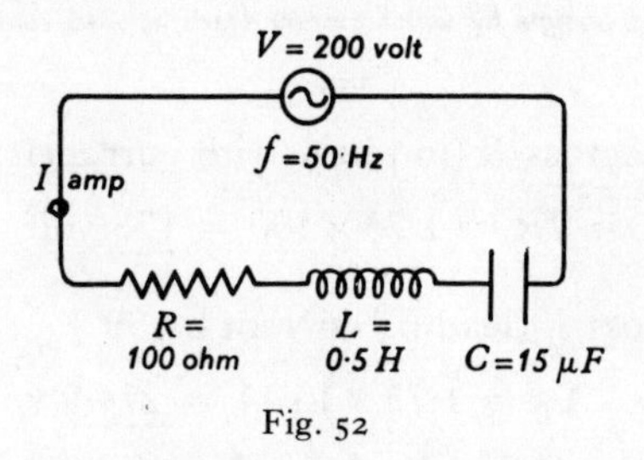

Fig. 52

The arrangement of the circuit is as shown in the diagram where R represents the resistance of the heater, L the inductance of the choke and C the capacitance of the capacitor.

(*a*) The impedance (Z) of the circuit is given by:

$$Z = \sqrt{R^2 + (X_L \sim X_C)^2}$$

where X_L = inductive reactance = $2\pi fL$ (f = mains frequency)

$$= 2\pi \times 50 \times 0{\cdot}5 = \underline{157{\cdot}1} \text{ ohm}$$

and $\quad X_C = \text{capacitative reactance} = \dfrac{1}{2\pi fC}$

$$= \frac{1}{2\pi \times 50 \times 15 \times 10^{-6}} = \underline{212{\cdot}2 \text{ ohm}}$$

Thus $\quad Z = \sqrt{100^2 + (212{\cdot}2 - 157{\cdot}1)^2} = \sqrt{100^2 + (55{\cdot}1)^2}$

$$= \underline{114{\cdot}2 \text{ ohm}}$$

(*b*) The current is given by Ohm's law. Thus,

$$I = \frac{V}{Z} = \frac{200}{114{\cdot}2} = \underline{1{\cdot}75 \text{ ampere}}$$

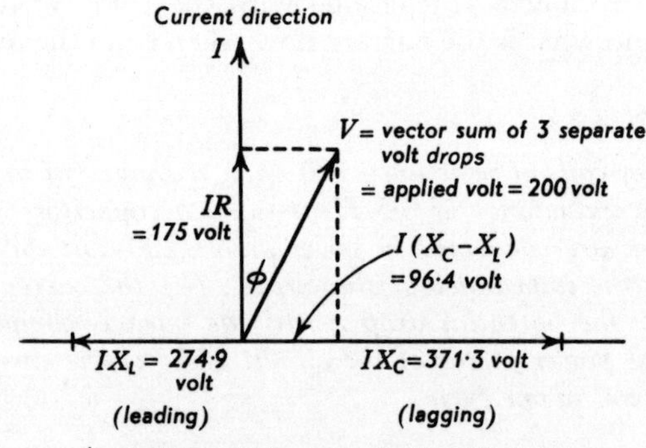

Fig. 53

(*c*) Potential drop across R (in phase with current)

$$= IR = 1{\cdot}75 \times 100 = \underline{175} \text{ volt}$$

Potential drop across L (leading current by 90°)

$$= IX_L = 1{\cdot}75 \times 157{\cdot}1 = \underline{274{\cdot}9} \text{ volt}$$

Potential drop across C (lagging behind current by 90°)

$$= IX_C = 1{\cdot}75 \times 212{\cdot}2 = \underline{371{\cdot}3} \text{ volt}$$

(*d*) It can be seen from the diagram above of the volts vectors that the current leads the applied voltage by an angle ϕ

$$= \tan^{-1}\left(\frac{X_C - X_L}{R}\right) = \tan^{-1}\left(\frac{55{\cdot}1}{100}\right) = 28°51'$$

The cosine of this angle gives the power factor of the circuit.

Thus, power factor $= \cos \phi = \cos 28°51' = \underline{0{\cdot}876}$

(*e*) The power in the heater

$= $ volts drop across $R \times$ circuit current

$= 175 \times 1{\cdot}75$

306·3 watts (Joules per sec.)

Hence energy converted by heater in one cycle (i.e. in $\frac{1}{50}$ sec.)

$$= 306{\cdot}3 \times \frac{1}{50} = \underline{6{\cdot}13} \text{ joule}$$

113 What change in the capacitance of the capacitor in the above problem would be needed to bring the current into phase with the supply volts and what would the current be under these conditions?

114 A resistance of 80 ohm, having a capacitor of capacitance 20 micro-farad in parallel with it, is connected across a 200-volt, 50-cycle supply mains. Calculate the total current taken from the mains, the impedance of the circuit, and the power factor for the complete circuit.

115 When connected to a 10-volt d.c. source a given coil passes 2 ampere. When the same coil is connected across an a.c. source of 10 volt (r.m.s.) and frequency 50 Hz, the current is only 0·5 ampere (r.m.s.). Explain this and find a value for the coil inductance. Find also the power factor of the coil in the a.c. circuit.

116 It is required to restore the power dissipated in the coil of the above problem to its value when the coil was connected across the d.c. source. Explain how a capacitor could be used to achieve this and calculate a value for its capacitance.

117 What must be the voltage of a 40-cycle a.c. mains supply to send a current of 2 ampere through a capacitor of capacitance 25 microfarad? If, using these mains, the current through the capacitor is to be reduced to 1 ampere by connecting a resistor in series with the capacitor, find the value of the required resistance.

118 An inductive coil has a self inductance of 0·1 H and carries a d.c. current of 2 A. What is the magnetic energy associated with the circuit?

If the coil has a resistance of 10 ohm, what current would it pass if connected across 100 V a.c. mains of frequency 50 Hz? Find also the power dissipated in the coil in these circumstances.

119 An inductive coil is connected across a 20-volt d.c. source through a depression key and an ammeter of negligible resistance. One second after closing the key the circuit current is observed to be 2 ampere, whilst after a further second the current recorded is 3 ampere. Explain this behaviour and calculate the inductance and resistance values of the coil.

What current would be passed by this coil if it were to be connected to 240-volt (r.m.s.) a.c. mains of frequency 50 Hz?

120 A series circuit containing an inductance L, a capacitance C and a resistance R is connected to an a.c. source of constant voltage V but variable frequency f. The circuit passes a maximum current when the supply frequency is 200 kHz, but when the frequency is reduced to 150 kHz, the current falls to half its maximum value. From this information determine the CR product for the circuit.

121 If, in the above problem, the value of C was 0·0005 μF, determine the values of L and R for the circuit, and so obtain the L/C ratio. Comment on the significance of the L/C and CR values for such a circuit.

122 An inductive coil is connected across an a.c. source whose frequency f and potential supply V are continuously variable. The following table gives the corresponding values of V and f as the current through the coil is kept constant at 10 mA at every stage.

f	100	150	200	250	300	350	Hz
V	2·24	3·16	4·12	5·15	6·08	7·06	r.m.s. volt

From a suitable plot of the results, obtain a value for the inductance of the coil. Also estimate a value for the resistance of the coil and comment on the result so obtained.

123 An a.c. source of 12 r.m.s. volt at a frequency of 50 Hz is placed across an inductive coil connected in series with a non-inductive 100-ohm resistor. When a valve voltmeter is placed in turn across the coil and resistor, readings of 7 and 8 r.m.s. volt respectively are recorded. Account for this and obtain values for the coil inductance and its resistance.

124 Two inductive coils, each of negligible resistance, are connected in series across a.c. mains of voltage (r.m.s.) 200 and frequency 50 Hz.

A current of 2 ampere (r.m.s.) is recorded, but this rises to 5 ampere (r.m.s.) when one of the coils is shorted out. Calculate the inductances of the two coils, and find the mains current when the two coils are connected in parallel across the same source.

125 An a.c. circuit is comprised of a non-inductive resistance in parallel with an inductive coil (the 'load'). This arrangement is connected across the mains supply with ammeters A_1, A_2 and A_3 in the supply circuit, the resistive branch and the inductive branch respectively. Draw a diagram of the arrangement and calculate the power factor of the 'load' if the ammeter readings are (in the above order) 3, 2 and 1·5 r.m.s. amp.

126 A choke coil of negligible resistance and a resistor are connected in series across a 200-volt a.c. supply and take a current of 2·5 amp at *p.f.* 0·8. If they are now connected in parallel across the same mains, determine the new power factor and the total current taken from the mains.

127 One branch of a parallel circuit consists of a resistor of 50 ohm in series with an inductance of 0·25 Henry, the other branch containing a capacitor of 10 microfarad. The complete circuit is connected across a 230-volt, 50-cycle supply. Find (*a*) the total current drawn from the mains, (*b*) the phase of the current.

Electronics and modern physics

Atomic physics

Some data and useful constants

unified atomic mass unit	u	$\approx$	$1{\cdot}6604 \times 10^{-27}$ kg
electron-volt	eV	$\approx$	$1{\cdot}6021 \times 10^{-19}$ J
mass of electron	m_e	$=$	$9{\cdot}1091 \times 10^{-31}$ kg
mass of proton	m_p	$=$	$1{\cdot}67252 \times 10^{-27}$ kg
mass of neutron	m_n	$=$	$1{\cdot}67482 \times 10^{-27}$ kg
charge of proton or electron	e	$=$	$\pm 1{\cdot}60210 \times 10^{-19}$ C
specific charge of electron	e/m_e	$=$	$-1{\cdot}758796 \times 10^{11}$ C kg^{-1}
Avogadro constant	L	$=$	$6{\cdot}02252 \times 10^{23}$ mol^{-1}
Planck constant	h	$=$	$6{\cdot}6256 \times 10^{-34}$ J s
speed of electromagnetic radiation	c	$=$	$2{\cdot}997925 \times 10^{8}$ m s^{-1}

Some work function voltages ϕ for use in photo-electric equation:

Caesium	1·90 V
Potassium	2·26 V

Sodium	2·46 V
Zinc	4·24 V
Tungsten	4·49 V

128 Define the terms: *amplification factor, stage gain* and *anode resistance* as they relate to a simple triode amplifier.

With such an amplifier a stage gain of 10 is obtained when the load resistance in the amplifier circuit is 10,000 ohm. When this resistance is doubled, the stage gain increases by 50 per cent. From these observations find a value for the anode resistance of the valve, also determine its amplification factor.

Worked example

129 *A triode valve has a mutual conductance of 2·5 mA per volt and an anode a.c. resistance of 20,000 ohm. What load resistance must be incorporated in the anode circuit to obtain a stage gain of 30?*

The stage gain

$$= \frac{\text{amplification factor } (\mu) \times \text{load resistance } (R)}{\text{anode a.c. resistance } (R_0) + \text{load resistance } (R)}$$

Hence $$30 = \frac{\mu R}{20{,}000 + R}$$

Now mutual conductance (g) $$= \frac{\text{amplification factor}}{\text{anode a.c. resistance}}$$

i.e. $$2{\cdot}5 \times 10^{-3} = \frac{\mu}{20{,}000}$$

or $$\mu = 50$$

Inserting this value for μ in the equation above we have

$$30 = \frac{50R}{20{,}000 + R}$$

from which $$R = \underline{30{,}000} \text{ ohm}$$

130 Describe how you would obtain the characteristic curves of a triode valve.

The table below gives data for two such curves for a triode valve. Draw these curves and obtain from them the amplification factor, mutual conductance and a.c. resistance of the valve.

Grid voltage	-9	$-7{\cdot}5$	-6	$-4{\cdot}5$	-3	$-1{\cdot}5$	0
Anode current in mA at anode voltage of 100	1·2	2·4	3·8	5·2	6·6	8·0	9·4
,, ,, ,, ,, 70	0	0·2	1·0	2·4	3·8	5·2	6·5

131 Describe the use of a triode valve for maintaining the electrical oscillations in a circuit containing a capacitance and an inductance.

The variable capacitor of a radio receiver is set at 0·00025 microfarad in order to receive signals of wavelength 434 metre. Calculate the inductance of the coil in the tuned circuit. (Velocity of electromagnetic waves $= 3 \times 10^8$ m s^{-1}.)

132 What is meant by the term *mutual conductance* of a valve?

A resistance-coupled amplifier provides a stage gain of 20 with a load resistance of 20,000 ohm, and a gain of 30 with a load resistance of 45,000 ohm. What is the mutual conductance of the circuit valve?

133 From experiments on cathode rays the specific charge of an electron is found to be $1{\cdot}76 \times 10^{11}$ C kg^{-1}. The electro-chemical equivalent of hydrogen is 0·0105 milligram per coulomb. Using this data, compare the mass of the electron with that of the hydrogen atom.

134 What voltage must be applied across the electrodes of a discharge tube to give electrons arriving at the anti-cathode a velocity of $4{\cdot}2 \times 10^7$ m s^{-1}? Assuming all the energy of the cathode stream to be converted into heat at the anti-cathode, what is the magnitude of the discharge current if the rate of production of heat energy at the anti-cathode is 50 J s^{-1}? (Specific charge of an electron $= 1{\cdot}76 \times 10^{11}$ C kg^{-1}.)

Worked example

135 *A p.d. of* 2000 *volt is applied across the electrodes of a low-pressure discharge tube. Calculate the maximum velocity acquired by the electrons in the cathode stream assuming the charge of the electron to be* $1{\cdot}59 \times 10^{-19}$ *coulomb and its mass to be* $9{\cdot}04 \times 10^{-31}$ *kg.*

When a charge of 1 coulomb passes between two points whose p.d. is 1 volt, 1 joule of energy is liberated.

Hence when a charge of $1{\cdot}59 \times 10^{-19}$ coulomb passes between two electrodes across which there is a p.d. of 200 volt, the energy liberated

$$= 1{\cdot}59 \times 10^{19} \times 2000$$
$$= 3{\cdot}18 \times 10^{16} \text{ joule}$$

The energy thus released goes to increase the kinetic energy of the electron, and if we consider an electron initially at rest on the cathode to acquire a velocity of v when arriving at the positive electrode, we have

$$\text{Increase in kinetic energy} = \tfrac{1}{2}m_e v^2$$

$$= \tfrac{1}{2} \times 9{\cdot}04 \times 10^{-31} \times v^2 \text{ joule}$$

Hence

$$\tfrac{1}{2} \times 9{\cdot}04 \times 10^{-31}\, v^2 = 3{\cdot}18 \times 10^{-16}$$

or

$$v = \sqrt{\frac{2 \times 3{\cdot}18 \times 10^{16}}{9{\cdot}04 \times 10^{-31}}}$$

$$= \underline{2{\cdot}65 \times 10^7 \text{ m s}^{-1}}$$

136 Obtain an expression for the force exerted on a fine beam of electrons travelling with a velocity v into a uniform magnetic field of flux density B directed perpendicular to the track of the electrons.

Such an electron stream is found to follow a track which forms part of a circle of radius 20·0 cm on entering a perpendicular magnetic field of intensity $7{\cdot}5 \times 10^{-4}$ Wb m^{-2}. Taking e/m_e as $1{\cdot}76 \times 10^{11}$ C kg^{-1}, find the velocity of the electron stream.

137 Two horizontal metal plates 3 cm long are positioned 0·8 cm apart in an evacuated enclosure. A fine beam of electrons enters the system travelling horizontally exactly mid-way between the plates and forms a spot on a vertical flurorescent screen situated in the enclosure 10 cm beyond the plates. A steady p.d. of 200 volt is now applied across the plates when it is observed that the spot is deflected through a distance of 2·0 cm. Describe the path of the electron as it passes between the plates, and taking the specific charge of the electron as $1{\cdot}76 \times 10^{11}$ C kg^{-1}, calculate the velocity of the electron stream as it enters the plates.

138 A stream of electrons, accelerated from rest through a potential of 900 V, enters a transverse magnetic field of intensity $0{\cdot}5 \times 10^{-3}$ Wb m^{-2}. Describe fully the path of the electrons as they travel through the magnetic zone. If the magnetic field has a circular cross-section of diameter 4·0 cm calculate the (approximate) subsequent deflection recorded on a vertical screen situated 12 cm from the central axis of the magnetic field.

139 The electrostatic deflection plates of a cathode-ray tube are positioned 2 cm apart and have a deflecting p.d. of 100 V across them.

If the subsequent deflection of the electron beam is neutralised by a co-terminal transverse magnetic field of 0·05 T, what is the velocity of the electron beam as it passes through the deflection system?

Worked example

140 *A fine beam of electrons, travelling horizontally with a velocity of* 3×10^7 *m s*$^{-1}$*, passes symmetrically between two identical metal plates positioned horizontally at a separation of* 1·0 *cm. An increasing potential difference is applied across the plates and it is found that when this is* 320 *V, the electron beam just strikes the end of one of the plates. If the length of the plates is* 4 *cm, calculate a value for the specific charge of the electron. (It is to be assumed that the system is contained in an evacuated enclosure.)*

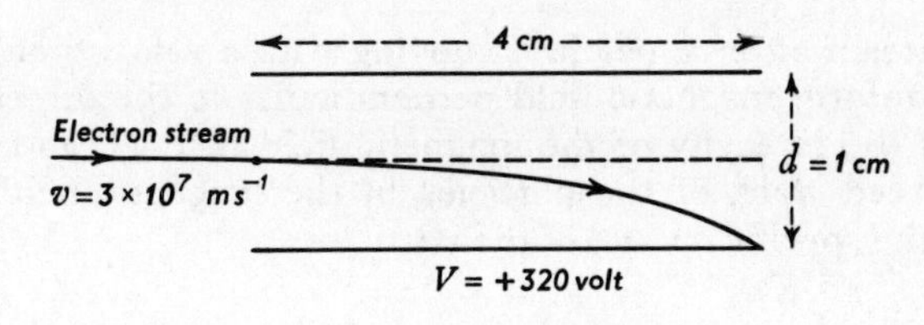

Fig. 54

The diagram shows the arrangement when the electron beam is just caught at the far edge of the lower plate. The field (E) between the plates $= \dfrac{V}{d} = \dfrac{320}{0{\cdot}01}$ volt per metre, and this produces a downward vertical force (F) on the electron (charge e) of Ee. Now if m_e is the mass of the electron, its downward acceleration (a) is given from $F = m_e a$.

Accordingly $$a = \frac{F}{m_e} = \frac{Ee}{m_e} = \frac{e}{m_e}\frac{320}{0{\cdot}01}\ \mathrm{m\ s^{-2}}.$$

Now the horizontal velocity of the electrons remains unchanged during their passage between the plates, and the time (t) for this part of their journey is $\dfrac{0{\cdot}04}{3 \times 10^7}$ seconds during which time the vertical deflection will be $\frac{1}{2}at^2$

or $$\frac{1}{2}\frac{e}{m_e}\frac{320}{0{\cdot}01}\left(\frac{0{\cdot}04}{3 \times 10^7}\right)^2 = 0{\cdot}005$$

(since just caught at the bottom plate edge).

Hence, the specific electronic charge $\left(= \dfrac{e}{m_e}\right)$

$$= \frac{0{\cdot}005 \times 2 \times 0{\cdot}01}{320} \times \left(\frac{3 \times 10^7}{0{\cdot}04}\right)^2$$

$$= \underline{1{\cdot}76 \times 10^{11}\ \text{C kg}^{-1}}$$

141 In the usual electron-gun arrangement of a cathode-ray tube the electrons are accelerated from rest to the annular anode by a potential of 1000 V. Calculate the velocity of the electrons as they emerge from the anode. If the distance between the cathode and anode is 5.0 cm and the fluorescent screen is positioned 10·0 cm beyond the anode, what time elapses between the electrons leaving the cathode and their arrival at the screen? (Specific charge of the electron $= 1{\cdot}759 \times 10^{11}$ C kg^{-1}.)

142 A stream of ionic particles moving with a velocity of 10^6 m s^{-1} enters a uniform magnetic field perpendicular to the direction of the stream. If the intensity of the magnetic field is 0·1 T and the radius of the curved track of the particles in the magnetic field is 0·2 m, calculate the specific charge of the particles.

143 Two horizontal metal plates are set 2 cm apart in an evacuated enclosure, the upper plate being maintained at a potential of 1000 V above the lower plate. Electrons are now caused to be released (at negligible velocity) from the lower plate. Find:

(*a*) the ratio of the electric to the gravitational forces acting on an electron on release,

(*b*) the kinetic energy of an electron on striking the upper plate.

(Take e as $1{\cdot}60 \times 10^{-19}$ C and m_e as $9{\cdot}1 \times 10^{-31}$ kg.)

144 A charged oil droplet is held stationary between two parallel plates when they have a p.d. of 3000 V across them. As a result of the acquisition of additional charges due to further ionization in the inter-plate zone, the droplet begins to move vertically. Find the ratio of the charges on the droplet in the two cases (i.e. before and after movement commences) if its motion can be halted by altering the applied p.d. by 1000 V.

145 Two plane parallel metal plates are set horizontally one immediately above the other at a separation of 2·0 cm, in air. A small charged oil drop, of mass $6{\cdot}53 \times 10^{-15}$ kg, can be held stationary between the plates when the upper plate is raised to a potential of 2000 V above the lower plate. How many electrons are attached to the drop? Find also the initial acceleration of the drop if (*a*) it acquired another electron, (*b*) the potential between the plates were suddenly reversed in the first situation. (Ignore the density of air in comparison with that of the oil.)

146 What do you understand by the terms *threshold wavelength* and *potential barrier* as applied to a photo-emissive surface?

A photo-cathodic antimony-caesium surface has a threshold wavelength of $6{\cdot}7 \times 10^{-7}$ m. What p.d., applied externally to the surface, will result in complete suppression of the photo-electrons from the surface?

147 A certain photo-cell is to be incorporated in a relay circuit which is triggered by a light beam incident on the cell face. Calculate the work function voltage of the cathode material if the arrangement is to discriminate against red light (i.e. wavelength greater than $6{\cdot}5 \times 10^{-7}$ m).

148 Describe the essential features of a modern X-ray tube and describe the energy changes occurring in the tube when powered to generate X-rays.

A 100-kV X-ray tube is fitted with a tungsten target. Calculate the wavelength of the X-radiation emitted by the target when the tube is in action and state whether you consider this to be an upper or lower limit when operating on this voltage. Take Planck's constant as $6{\cdot}63 \times 10^{-34}$ Js and the electronic charge as $1{\cdot}60 \times 10^{-19}$ C.

149 Give the meaning of the term *work function* as applied to the photo-electric emission of electrons from the surface of a metal.

Calculate the work function voltage of a metal surface for which the maximum release energy of the electrons is $0{\cdot}82 \times 10^{-19}$ J when illuminated with radiation of frequency $6{\cdot}17 \times 10^{14}$ Hz. Take the electronic charge as $1{\cdot}60 \times 10^{-19}$ C and Planck's constant as $6{\cdot}63 \times 10^{-34}$ Js.

150 In an experiment on photo-electric emission, it was found that when radiation of wavelength $4{\cdot}5 \times 10^{-7}$ m was incident on a metal surface the maximum energy of the electrons emitted was $0{\cdot}42 \times 10^{-19}$ J but was $2{\cdot}63 \times 10^{-19}$ J when radiation of wavelength $3{\cdot}0 \times 10^{-7}$ m was used. Explain these observations and obtain a value of Planck's constant from them.

151 Electrons are only emitted from a metal surface if the frequency of the incident light exceeds a certain value. Explain why this is and calculate the minimum frequency for which electrons can be emitted from a metallic surface from which the maximum kinetic energy of the released electrons is $1{\cdot}50 \times 10^{-19}$ J when the surface is illuminated by radiation of frequency $4{\cdot}57 \times 10^{14}$ Hz. Take Planck's constant as $6{\cdot}63 \times 10^{-34}$ J s.

152 State the expression giving the equivalence of mass and energy, and, using the data given on p. 171, obtain the equivalent energy value (*a*) in electron-volts, (*b*) in Joules, of the atomic mass unit.

153 Define *half-life* and *decay constant* of a radioactive nuclide. How are these quantities related?

An ionizing chamber is placed over an extension electrode from a pulse electroscope and thoron gas from thorium X source is forced into the chamber. The pulse rate recorded by the electroscope after the injection of the thoron gas was $0{\cdot}65\ s^{-1}$ and 20 seconds later it had fallen to $0{\cdot}50\ s^{-1}$. Calculate the half life of the thoron gas from these observations.

154 What mass, in grammes, of an original mass of 4 grammes of radium would remain after a lapse of 200 years given that the half-life period for radium in 1600 years?

155 A Geiger-Muller counting tube with a window area of 4 cm^2 is clamped in a vertical position so that the window is 20 cm immediately above a small β-source. When so placed an average count of 3000 ionizing events per minute is recorded by a Scaler unit connected to the Geiger tube. Estimate the strength of the source, in Curies, being given that the Curie is defined as the activity of a source which undergoes $3{\cdot}7 \times 10^{10}$ disintegrations per second.

State two main assumptions implied in this method of estimating source strengths.

156 An experiment on the absorption of β-rays was conducted by inserting aluminium foils between a β-source and the detecting face of a Geiger-Muller tube. The readings of a ratemeter connected to the Geiger-Muller tube were found to be as set out below for the stated number of aluminium foils.

Number of foils	0	1	2	3	4	5	6
Count rate (per minute)	9130	2820	795	280	95	35	30

Show that the attenuation of the β-rays follows closely a logarithmic plot, and from your graph find the limiting range of the β-particles in aluminium (the thickness of the aluminium foils—of equal thickness—was 0·06 cm).

Find also the linear-absorption coefficient for β-rays in aluminium.

157 What is the law of attenuation of γ-rays by distance? How would you verify your statement experimentally?

The detecting area of a Geiger-Muller is positioned 10 cm away from a small 5 μCi γ-source when an average count rate of 600 s^{-1} was recorded by a ratemeter connected to the tube. The same tube was now placed with its detecting area 15 cm from another γ-source when the average count rate recorded was 400 s^{-1}. What was the strength of the second source? State any assumptions made in your calculation.

158 A stage in the disintegration of uranium 238 is represented by the following reaction:

$$_{92}U^{238} \longrightarrow {}_{90}Th^{234} + {}_{2}He^{4}$$

Explain the terms of this expression and using the result of question 152 (or otherwise) and the data given below,

(*a*) show that the reaction is possible,

(*b*) calculate the total energy (in Joules) released in the disintegration,

(*c*) obtain the value of the initial velocity of the alpha particle after the disintegration process assuming the nucleus to be initially at rest.

Mass of U^{238} = 238·1249 a.m.u.

" " Th^{234} = 234·1165 "

" " He^{4} = 4·0039 "

(Further data available on p. 171.)

ANSWERS TO NUMERICAL QUESTIONS

Mechanics and Hydrostatics

1 35 kg. **2** 1·25 cm from the centre of the original sheet measured along the line of symmetry on the side opposite to the hole in the sheet. **4** Long side makes an angle of 14° 2′ with the vertical. **5** 147 N (on left rod), 245 N (on right rod), vertically; 139·5 N 20° 33′ above the horizontal; 129·3 N. **6** 133·3 g. **7** 24·50 g; left arm: right arm = 0·981 : 1. **8** Arms of the balance are equal in length, fault due to right-hand pan being 0·12 g heavier than left-hand pan. **9** 2 mm below centre knife edge. **10** 5·5 kg; 25 kg. **11** $1·51 \times 10^6$ Nm^{-2}. **12** 0·49 N; 0·58. **14** 4·19 m. **15** 0·565 N. **16** 0·33. **17** 2·2. **19** 1·23; 0·92. **20** 0·7. **21** 0·68. **22** 0·62365 N. **23** 3 vols A to 2 vols B. **25** 116 g. **26** 71·25 g; 63·75 g. **28** 2.48. **29** $4·41 \times 10^5$ N at a point along the mid-axis $4\frac{3}{8}$ m from the top. **30** 0·47 N; 4·7 N; 1·88 N. **31** 2 min. **32** 5·66 ms^{-2}. **33** 952·6 m. **35** 95·6 m; 220·9 m; 8·84 s; 63·77 m. **36** 277·1 m; 82·04 ms^{-1}. **37** 1 : 2. **38** 100 m. **39** 5325 N. **40** 63 m. **41** 1087·5 N. **43** 4·8 N; 1·84 cm lower. **44** 1·106 s; 0·817 N; 0·734 N. **45** 196·4 N. **46** 9898 N. **47** 1·67 kW. **48** Loss of momentum is 0·0224 kg m s^{-1}; kinetic energy lost 0·049 J. **49** 25 J; 1000 N. **50** 82·6 m s^{-1}. **52** $3·87 \times 10^{-6}$ °C. **53** 0·82. **54** 64 per cent loss; 1·6 U m s^{-1}. **55** $M/m = 2$; $e = 1$. **56** 5·75 s; 911·1 cm. **57** The larger sphere proceeds with a velocity of 10 cm s^{-1} in a direction perpendicular to the line of centres, the smaller one rebounding with a velocity of 20 cm s^{-1} at an angle of 30° with the line of centres but on the side opposite to the approach direction. **58** 4·08 s; 70·62 m. **59** 31·4 cm s^{-1}; 0·0197 N. **60** 7° 10′. **61** 45 m. **62** 151 revs per min. **64** 14 m s^{-1}. **65** 6·3 revs per sec. **68** 9·99 cm; 3·14 s. **70** 9·845 m s^{-2}. **71** (i) 2·005 s; (ii) 0·5461 m s^{-1} (iii) 0·5054 N. **73** 0·8. **74** $2·47 \times 10^{-5}$ J; $1·85 \times 10^{-5}$ J. **75** 0·63 s. **76** 7·07 cm. **77** $T = 2\pi\sqrt{\frac{MV}{\gamma p A^2}}$; 0·168 s. **78** 0·174 s. **81** 3×10^{-3} J. **82** 3·75 cm. **83** Relative times of loop: disc : sphere : sliding body are 1·41 : 1·22 : 1·18 : 1. **84** 1 : 2 : 6. **85** 9·37 m s^{-2}; 9·84 m s^{-2}. **87** 0·577; 196 rad s^{-2}. **88** 0·904 s; 1·107 s. **89** 0·785 N m. **90** 4·88 N; $9·62 \times 10^{-7}$ J. **91** 2.48×10^{-3} kg m^2. **92** 0·000941 kg m^2. **93** (*a*) 197·4 J; (*b*) 0·105 N m; (*c*) 60 s. **94** 1·75 s. **95** $1·096 \times 10^{-3}$ kg m^2. **96** 0·913 : 1. **97** 110·36 cm. **98** 0·78 s. **100** 147·3 cm. **101** 1·728 s. **102** 9·80 m s^{-2}; $8·57 \times 10^{-3}$ kg m^2. **103** 0·013 per cent. **104** 1·524 s; 45·1 g. **105** When the C.G. of the rod is close to a position mid-way between the knife edges. **106** 0·165. **107** 0·73 s; 178·6 cm s^{-1}.

Properties of Matter

1 2×10^{11} N m^{-2}. **2** 8·4 kg. **3** 2:9. **4** 2·85 m. **6** 0·186 J. **7** 1 kg. **8** $9{\cdot}56 \times 10^3$ N. **10** 0·33. **11** $7{\cdot}85 \times 10^{10}$ N m^{-2}. **12** $1{\cdot}3 \times 10^{10}$ N m^{-2}. **13** $1{\cdot}12 \times 10^{10}$ N m^{-2}. **14** 0·294 m. **15** 0·072 N m^{-1}. **16** $2{\cdot}6 \times 10^{-4}$ m. **17** 0·041 m. **19** 850 kg m^{-3}. **20** 0·029 N m^{-1}. **22** 3·2 N m^{-2}; 0·0025 J. **24** $1{\cdot}0413 \times 10^5$ N m^{-2}. **26** 5·68 N. **27** 0·0207 N m^{-1}. **28** 54·7 rad s^{-1}. **29** 0·0012 N m^{-2} s. **30** 0·00172 N m^{-2} s. **31** $2{\cdot}5 \times 10^{-4}$ N. **32** $2{\cdot}73 \times 10^{-2}$ m s^{-1}; 1 N m^{-2} s. **33** $7{\cdot}1 \times 10^{-4}$ m. **35** $8{\cdot}92 \times 10^{-7}$ m. **36** $5{\cdot}68 \times 10^{-6}$ m. **37** (*a*) $1{\cdot}04 \times 10^{-6}$ m; (*b*) $1{\cdot}61 \times 10^{-19}$ C. **38** 215 min. **39** 8·54 s. **41** 6×10^{-7} cm. **43** 1·0016 s. **44** $2{\cdot}279 \times 10^8$ km. **45** $3{\cdot}83 \times 10^5$ km. **46** 9·815 m s^{-2}. **47** 0·9995:1. **48** 6·3′. **49** $7{\cdot}93 \times 10^3$ m s^{-1}. **50** 7920 m s^{-1}. **51** 5076 s. **52** 4267 km. **53** $35{\cdot}95 \times 10^6$ m; $3{\cdot}078 \times 10^3$ m s^{-1}.

54 $1{\cdot}12 \times 10^4$ m s^{-1}. **55** (*a*) all but helium (*b*) none. **56** $v_e \dfrac{dM}{dt}; \dfrac{v_e^2}{2}\dfrac{dM}{dt}$. **57** $4{\cdot}29 \times 10^3$ m s^{-1}. **58** (i) LT^{-2}; (ii) MLT^{-2}; (iii) MT^{-2}; (iv) $ML^{-1}T^{-1}$; (v) $M^{-1}L^3T^{-2}$. **59** $f = k\sqrt{\dfrac{E}{\rho l^2}}$ where E = Young's Modulus, ρ = density and l = length of prong. **61** 75 s. **62** $v = k\sqrt{\dfrac{Fl}{m}}$. **63** 23·5 cm s^{-1}.

Heat

4 889°C. **5** 22·2°C. **6** 45·3°C. **7** (*a*) 67°C; (*b*) 44°C. **8** Lengths of rods: brass = 10 cm, iron = 15 cm. **9** 139·1°C. **11** 0·022 per cent. **12** Time lost = 60·5 s. **13** 3454 N. **14** $13\frac{1}{3}$ s. **15** 22·3°C. **16** (*a*) 0·9 per cent increase; (*b*) 0·0016 per cent decrease. 17 140·4°C. **18** 0·805 cm^3. **20** 0·067 cm^3. **21** $\frac{2}{3}$ cm^3. **22** 74·893 cm; 74·622 cm. **23** 5·22 g; 0·0006 K^{-1}. **24** 96·86 g. **25** 0·961 g cm^{-3}. **26** 80·74 cm. **27** 3·5 cm. **29** 75·7 cm. **30** 7·08 m. **32** 49. **33** 63·5 m^3. **34** 55·07°C. **35** 8·314 J mol^{-1} K^{-1}. **36** 0·248. **38** 0·006°C. **39** (*a*) $3{\cdot}18 \times 10^{-4}$ kg; (*b*) 0·822 atmos. **40** 342·6 J. **41** 1925 J kg^{-1} K^{-1}; 66 J K^{-1}. **42** 2100 J kg^{-1} K^{-1}. **43** 19°C. **44** (i) exothermic reaction takes place with the liberation of 25200 J of heat energy; specific heat capacity of liquid = 2800 J kg^{-1} K^{-1}. **45** The same. **47** 41·2 J K^{-1}. **48** 634 J kg^{-1} K^{-1}. **49** 40°C; 35 min. **50** 24°C. **51** 1·34 kg. **52** $8{\cdot}51 \times 10^5$ J kg^{-1}. **53** 357 J kg^{-1} K^{-1}. **54** 22,680 J kg^{-1}. **56** 86·8 J kg^{-1} K^{-1}. **57** 413·6 J kg^{-1} K^{-1}. **58** 15·74 cm.

59 2083 J. **62** (*a*) $\frac{1}{4}$ atmos; (*b*) 0·099 atmos. **63** 200·5 cm of mercury; 122·9°C. **65** 0·432 the way down the cylinder. **66** Molar heat capacities (*a*) at constant volume = 20·18 J mol^{-1} K^{-1}; (*b*) at constant pressure = 28·44 J mol^{-1} K^{-1}. **67** (*a*) 9·52 atmos; 279·1°C; (*b*) 763·4°C. **68** 9·05 J increase. **70** (*a*) 150·95 cm of mercury; (*b*) 38·0 cm of mercury. **71** 66·4 cm. **72** Pressure must be increased to 118·5 cm of mercury. **73** 42·5 cm of mercury. **75** 9 mm of mercury; space becomes saturated during first compression. **76** (*a*) 79·7 per cent; (*b*) 100 per cent. **77** 40·2 per cent. **78** 1·2103 g. **79** 13.2 mm of mercury. **85** (*a*) $3{\cdot}57 \times 10^{12}$; (*b*) $3{\cdot}01 \times 10^{12}$. **86** 3295°C. **88** $4{\cdot}834 \times 10^{2}$ m s^{-1}. **90** $4{\cdot}643 \times 10^{2}$ m s^{-1}. **91** 194·9°C. **92** 2·45 s. **93** 7·4°C. **94** 0·0036°C s^{-1}. **95** 0·025 J cm^{-2} s^{-1}. **96** 8·48 g. **98** 7·5°C; 382·5°C. **99** (*a*) 27·3°C; (*b*) $6{\cdot}19 \times 10^{6}$ J. **100** 98 s. **101** 36 min. 48 sec. **102** 0·65°C s^{-1}. **104** 16·7°C. **105** 0·159 Wm^{-1} K^{-1}. **110** 2·8°C min^{-1}. **112** 5457°C.

Light

2 4 cm. **4** 20 cm. **5** (*a*) virtual, $6\frac{2}{3}$ cm behind the mirror; (*b*) real, 20 cm in front of the mirror. **6** 10·0 cm. **8** 61·8 cm. **9** 15 cm. **10** 1·34. **11** 1·49. **12** 62° 44′. **13** 1·33. **14** 1·35. **16** 13·6 cm. **17** 2979 m. **18** 6·78 cm. **19** 37° 12′; 48° 36′. **20** 29° 30′. **21** 83° 38′. **23** 1·63. **24** 1·45. **25** 1·71. **26** 1·51; 8. **28** 576 km hr^{-1}. **29** 15 cm. **30** 7·5 cm; 40 cm. **31** 24 cm; lens is 7 cm from the end of the tube nearer the screen. **33** 10 or 60 cm. **34** (*a*) virtual, 10 cm in front of A; (*b*) real, at centre of B; (*c*) real, 15 cm behind B; (*d*) real, 30 cm behind B. **35** At all points in front of the lens combination except for the range of distance between $3f/2$ and $4f/3$ from the first lens. **36** 25 cm; $45\frac{1}{3}$ cm. **37** 37·5 cm above the liquid surface. **38** 8 cm. **39** 1·50. **41** 1·33. **42** (i) 1·54; (ii) 19·7 cm. **43** Real, 30 cm from the centre of the sphere on the side opposite to the object. **44** (i) 2·5 cm; (ii) 8·75 cm—both distances being measured from the appropriate surface at the point where it is intersected by the viewing axis. **45** 20 cm. **46** 30 cm; 1·50. **48** 1·414. **49** 1·0 mm. **50** Real, 6·39 cm behind concave lens, with relative size of 2/77. **51** (*a*) 593 cm or 20·3 cm from the object; (*b*) 345 cm from the object. **52** 85 cm. **53** 0·0155; 4·8′. **54** 8°; 4·8′. **55** 4°; 2·57°. **56** 1·54 cm. **58** Concave flint glass lens of focal length 57·6 cm, convex crown glass of focal length 36·5 cm. **62** 45° 42′ measured from the axis of the bow through the eye of the observer. **63** 230° 6′, 232° 44′; 2° 38′. **65** Concave, of focal length 5 m. **67** Concave lenses of power 4 dioptre; infinity. **68** (*a*) convex lenses of focal length 37·5 cm giving a range of vision from 25·0 to 31·6 cm; (*b*) concave spectacles of focal length 200 cm giving a range of vision from 120 cm to infinity. **69** $4\frac{2}{7}$ cm from the lens; 7. **71** $37\frac{2}{3}$. **72** $1\frac{6}{89}$ cm from objective; eyepiece should be moved in by $\frac{1}{6}$ cm. **74** Objective, 20 cm; eyepiece, 5 cm. **75** Eyepiece must be moved in by $\frac{1}{6}$ cm. **78** 37·4′. **81** 9·8 cm. **84** 120; (i) halved; (ii) unaffected; (iii) decreased to one quarter of original value. **85** $2{\cdot}00 \times 10^8$ m s^{-1}. **86** 16 min. 33·3 sec. **88** Greatest interval is 42 hr 29 min 15·2 sec, least interval is 42 hr 28 min 44·8 sec. **89** 2250 rev per min. **90** (*a*) 4500 rev per min; (*b*) 11,250 rev per min. **92** 47·75 rev per sec. **94** 9:16; $33\frac{1}{3}$ cm from weaker lamp between the two lamps, or 100 cm from the weaker lamp on the side remote from 20 cd lamp. **95** 80 cm from the screen on same side as 20 cd lamp. **97** 46·2 cm. **98** 0·51. **99** 1·035 m vertically above the lamp. **100** 0·48. **101** 125:64. **102** 281 cd. **104** 0·0442 mm. **105** $5{\cdot}89 \times 10^{-7}$ m. **106** 0·203 mm. **107** 0·0227 mm. **108** $5{\cdot}88 \times 10^{-7}$ m. **109** 1·34. **110** $2{\cdot}38 \times 10^{-5}$ cm. **111** 7500. **112** (*a*) 3; (*b*) 3′. **115** $5{\cdot}9 \times 10^{-7}$ m; 3. **116** $1{\cdot}704 \times 10^{-10}$ m. **120** 0·000164 m.

Sound

3 Sinusoidal wave of amplitude 10^{-6} m and wavelength 0·68 m travelling in the positive direction of x with a velocity of 340 m s^{-1}. **4** 252 Hz. **5** 128, 144, 160, 170·6, 192, 213·3, 240, 256. **7** 508 Hz; 512 Hz. **8** 250 Hz; 256 Hz. **9** 83 db. **11** 12·55. **13** 342 m s^{-1}. **14** $3{\cdot}31 \times 10^2$ m s^{-1}. **15** $4{\cdot}96 \times 10^{-5}$ cm^3 contraction. **17** 462 m. **18** 333·3 m s^{-1}. **19** 344 m s^{-1}. **20** 333 m s^{-1}; 350 m. **21** 344·3 m s^{-1}. **22** 5:4. **24** 530 Hz. **25** 0·214 m. **26** 318·2 m s^{-1}. **28** 339 m s^{-1}. **29** 332·1 m s^{-1}; 1·14 cm. **31** 2698 Hz. **32** $7{\cdot}57 \times 10^{10}$ Nm^{-2}. **33** 1·4. **34** 70·2 cm. **35** 388 Hz. **37** 248 Hz. **38** 10×10^{10} N m^{-2}. **40** 250 Hz. **41** Load must be reduced to 225 g. **44** 2:3; original length of A = 30 cm, of B = 20 cm. **45** (i) $T_2 - T_1 = 0{\cdot}24$ kg; (ii) $l_1 - l_2 = 0{\cdot}99$ cm. **46** 8·61. **48** 378 Hz. **49** 96 Hz. **50** 1266 Hz. **51** 3720 Hz. **52** (*a*) 600 Hz; (*b*) 625 Hz; (*c*) 733 Hz. **53** 2·72 m s^{-1}; no, B hears beats of frequency 7·8 s^{-1}. **55** 436 to 586 Hz. **56** 933 Hz; 1077 Hz. **57** Falls from 1466 Hz to 759 Hz. **58** 332 Hz. **59** 0·031 Å. **60** Velocity of recession = $6{\cdot}91 \times 10^4$ m s^{-1}. **61** Star receding with speed of $2{\cdot}406 \times 10^4$ m s^{-1}; star rotating with outer 'limb' velocity of $7{\cdot}4 \times 10^3$ m s^{-1}. **62** $7{\cdot}0 \times 10^8$ m.

Electrostatics

1 1 nC; along the line of centres at a point between them 8·29 cm from the centre of the smaller sphere. **3** 233 V. **4** 0, 720 V, 240 V. **5** 1656 V, 360 V, −360 V; $21\frac{3}{7}$ cm from A's centre along line of centres.

6 $1{\cdot}5 \times 10^6$ V. **7** $8{\cdot}84 \times 10^{-10}$ C m^{-2}. **8** 3. **9** (*a*) 6 μF; (*b*) $\frac{2}{3}$ μF; 2 in series with the other in parallel with them. **10** 50 V. **11** (*a*) $1{\cdot}2 \times 10^{-4}$ C; (*b*) 60, 40, 20 V; (*c*) $4{\cdot}8 \times 10^{-3}$ J. **12** $4{\cdot}42 \times 10^{-8}$ J; (i) Energy is halved, (ii) energy is doubled. **14** $3{\cdot}6 \times 10^{-3}$ J; 10^{-2} J. **15** (*a*) $28\frac{4}{7}$ V; (*b*) $0{\cdot}643^{-3}$ J. **16** 10^{-9} F; energy at breakdown $= 3{\cdot}125 \times 10^{-3}$ J, maximum initial energy $= 0{\cdot}625 \times 10^3$ J. **17** Potentials are respectively: 483·9, 322·6 and 193·5 V; each carries the same charge of $9{\cdot}68 \times 10^{-4}$ C; total energy in series = 0·494 J, in parallel = 5 J. **18** 2·5. **19** 6. **20** (*a*) $8{\cdot}80 \times 10^{-12}$ F m^{-1}; (*b*) $6{\cdot}6 \times 10^{-12}$ F. **21** 0·98 μF. **22** $\left(\frac{n-1}{n}\right)C_1$. **23** 1·6 μF. **24** $58\frac{1}{3}$ V. **25** $-27\frac{3}{31}$, $2\frac{28}{31}$, $62\frac{28}{31}$ μC; potential at B is $13\frac{17}{31}$ V above that of A, potential at C = potential at B, potential at D (= potential at E) is $12\frac{18}{31}$ V above that of F, potentials at A and F the same. **26** (*a*) $2{\cdot}65 \times 10^{-8}$ C; (*b*) $9{\cdot}28 \times 10^{-8}$ C. **27** Potentials of the spheres are respectively: 3780, 3321 and 2430 V; potentials of inner and outer spheres fall to 405 and 675 V respectively; $-6{\cdot}75$ nC. **28** $5{\cdot}56 \times 10^{-13}$ A. **30** 8·66 μF. **31** An additional mass of 0·142 g is required. **32** $2{\cdot}21 \times 10^{-5}$ N. **33** $1{\cdot}83 \times 10^{-11}$ F. **34** 2·42. **35** 44·8 nC. **36** $1{\cdot}23 \times 10^{-15}$ kg.

Current Electricity

1 1·33:1. **2** 1·5 volt; 2 ohm. **3** 4 per cent below true value. **5** (*a*) 3 amp; (*b*) 0·399 amp; 2 parallel rows of 15 cells each, 3·43 amp. **6** 1000 ohm; 500 ohm. **7** $\frac{3}{4}$ ohm. **8** 1·25:1. **9** (*a*) 16·2 ohm; (*b*) 10^{-7} ohm. **10** Current through, A = 0·77 amp, B = 0·23 amp. **11** 24·54 cm; assumed value of p.d. is 4.1 per cent less than true value. **12** 0·00608 K^{-1}; 32·9°C. **14** 1·77 volt. **15** 1000 ohm; $\frac{1}{60}$ ohm. **16** 16·7 volt. **17** 2 ohm; 1 amp; 26 volt. **18** 19·5 milliamp; 2·39 ohm. **19** $1{\cdot}17 \times 10^{-7}$ Ω m. **20** 8 ohm; 1:2. **21** 2·0 cm at the 0 cm end of the wire, 0·5 cm at the 100 cm end. **22** 21·75 and 9·05 ohm. **24** 1·45 volt. **26** Anomalous readings due to zero error (on account of end resistance) of 0·5 cm of wire; 25 ohm. **27** 3·2 ohm; 66·2 ohm. **28** Ammeter over-reads by 0·1 amp. **29** 73·3 cm. **30** 149 volt. **31** Arrange component parts of the circuit in series with a combined resistance of 1257 ohm between the boxes A and B: connect one terminal of the Weston cell to the common connection of the two boxes, the other terminal being connected to a key tapping on the potentiometer wire. Possible balance positions are:

26·7 cm of the wire with 608 ohm in A, 649 ohm in B
60·0 ,, ,, ,, ,, ,, 609 ,, ,, ,, 648 ,, ,, ,,
93·3 ,, ,, ,, ,, ,, 610 ,, ,, ,, 647 ,, ,, ,,

the box B being connected next to the wire.

33 1·875 millivolt. **34** (i) 2·31 millivolt, (ii) 332°C, (iii) 209·4°C or 454·6°C. **35** 1·97 millivolt. **36** 0·0395 g of copper; 0·0407 g of zinc. **37** Ammeter over-reads by 0·005 amp. **38** $6{\cdot}024 \times 10^{23}$. **39** 1·09 volt. **41** 1·54 volt. **42** 0·91 amp; 20 ampere-hr. **43** 5 :3; resistance of A = 5·08 ohm, of B = 3·05 ohm. **44** (*a*) 2 :5; (*b*) 5 :2; 4 :49. **46** 24·5 ohm; 12·5 per cent loss. **47** 70 per cent; efficiency increased to 95·2 per cent. **48** 14·2°C. **50** 45°C. **51** 3·39 amp. **52** 0·7 sec (approx), assumptions made: resistivity and specific heat capacity of copper constant, no surface heat losses from wire. **54** 900 amp. **56** Force of attraction of $2{\cdot}5 \times 10^{-4}$ N m^{-1} **57** 36 N. **58** (*a*) $1{\cdot}5 \times 10^{-3}$ N; (*b*) Nil; (*c*) $0{\cdot}75 \times 10^{-3}$ N. **59** At a point between the wires distant 4 cm from wire A on the line joining A to B. At a point distant 40 cm from A on the side remote from B along the line joining A to B. **60** $1{\cdot}8 \times 10^{-5}$ Wb m^{-2}. **61** 10^{-5} N; $1{\cdot}387 \times 10^{-5}$ J. **62** 4×10^{-5} N. **63** 0·184 Wb m^{-2}. **64** 0·0126 N. **65** Plane of coil perpendicular to meridian with a current of 0·143 amp circulating in an anticlockwise direction on looking along the meridian northwards to the face of the coil. **66** 1·6 amp. **67** (*a*) $3{\cdot}35 \times 10^{-4}$ Wb m^{-2}; (*b*) $1{\cdot}675 \times 10^{-4}$ Wb m^{-2} in reverse direction from (*a*); (*c*) $3{\cdot}263 \times 10^{-4}$ Wb m^{-2} at an angle of 22° 38′ with original direction. **68** Arrange coils coaxially at a separation equal to coil radius and with currents circulating in same sense through both coils; $1{\cdot}96 \times 10^{-5}$ Wb m^{-2}. **69** 0·058 amp. **70** 318·3. **71** $3{\cdot}75 \times 10^{-5}$ N m. **72** 0·015 N; $2{\cdot}5 \times 10^{-6}$ N m per degree. **73** (*a*) Place a shunt of resistance 0·01505 ohm in parallel with the galvanometer; (*b*) connect a resistance of 661·7 ohm in series with the galvanometer. **74** 22·9°. **75** 2·1 μF. **76** 21·9 divisions. **77** 2·89 megohm. **78** 0·092; 31·4 divisions. **79** $6{\cdot}28 \times 10^{-3}$ Wb; Flux reduced to $0{\cdot}7 \times 10^{-3}$ Wb. **80** 0·43 millivolt. **81** 0·236 volt. **83** 0·133 volt. **84** (*a*) 0·474 volt; (*b*) 0·237 volt. **85** 0·102 T (or Wb m^{-2}). **86** 67° 45′. **87** $3{\cdot}18 \times 10^{-2}$ Wb m^{-2}. **88** 0·10 Wb m^{-2}. **89** 4·21 $\times 10^{-4}$ H; $6{\cdot}32 \times 10^{-6}$ Wb. **90** 19,800 turns; 0·02 amp. **92** 6·25 × $\times 10^{-3}$ H. **93** 2828 volt. **94** 0·14 sec; 2000 volt. **96** (*a*) 5 amp; (*b*) 2 amp per sec. **97** 1000. **98** 5 amp per sec; 16 joule. **99** 0·001675 henry; 0·042 volt. **101** 167·5 rev per min. **103** 9·9 amp. **104** $3\sqrt{5}$ r.m.s. amp. **105** 4·69 amp. **106** 62·8 ohm; 2·86 amp. **107** (*a*) 2116 watt; (*b*) 4232 watt; (*c*) 42·4 joule. **108** (*a*) 0·15 r.m.s. amp; (*b*) 95·4 r.m.s. volt; (*c*) 4·5 watt. **109** 0·00188 henry. **110** 0·135 μF; 5 milliamp r.m.s. **111** 400 ohm, 2·21 henry; 3·45 μF, 0·125 r.m.s. amp. **113** Capacitance should be increased to 5·27 μF; 2 amp. **114** 2·8 amp; 71·4 ohm; 0·89. **115** 0·062 henry; 0·25. **116** Connect a capacitor, of capacitance $1{\cdot}69 \times 10^{-4}$F in series with the coil. **117** 318·4 volt; 275·7 ohm. **118** 0·2 joule; 3·03 r.m.s. amp; 91·8 watt. **119** 7·21 henry, 5 ohm; 0·106 r.m.s. amp. **120** $2{\cdot}68 \times 10^{-7}$ ohm-farad. **121**

1·27 milli-henry; 53·6 kilo-ohm; 2·54 × 10^{6} henry-farad^{-1}. **122** 0·32 henry; 100 ohm. **123** 0·267 henry; 25 ohm. **124** 0·1274, 0·1911 henry; 8·33 r.m.s. amp. **125** 0·46. **126** 0·6; 5·21 r.m.s. amp. **127** (*a*) 1·90 r.m.s. amp; (*b*) current lags by 45° 38′. **128** 20,000 ohm; 30. **130** 10; 9 × 10^{-2} amp per volt; 11,111 ohm. **131** 212 micro-henry. **132** 1·67 milliamp per volt. **133** 1:1848. **134** 5010 volt; 9·91 milliamp. **136** 2·64 × 10^{7} m s^{-1}. **137** 2·755 × 10^{7} m s^{-1}. **138** Circular path of radius 0·20 m; 0·024 m. **139** 10^{5} m s^{-1}. **141** 1·875 × 10^{7} m s^{-1}; 1·066 × 10^{-} s. **142** 5 × 10^{7} C kg^{-1}. **143** (*a*) 9·0 × 10^{14}: 1; (*b*) 1·60 × 10^{-16} joule. **144** 2:3. **145** 4; (*a*) 2·45 m s^{-2} vertically upwards; (*b*) 19·6 m s^{-2} vertically downwards. **146** p.d. > 1·85 volt. **147** 1·91 volt. **148** 1·24 × 10^{-11} m; lower limit. **149** 2·04 volt. **150** 6·63 × 10^{-34} Js. **151** 2·31 × 10^{14} Hz. **152** 931·46 MeV; 1·4923 × $^{-10}$ J. **153** 52·8 s. **154** 3·67 g. **155** 1·7 × 10^{-6} Ci. **156** Range = 0·28 cm; linear absorption coefficient, defined from $\log_e \dfrac{I}{I_0} = -\mu x$, I being the incident intensity at a section distant x from the initial boundary (intensity I_0), is 1·93 × 10^{3} m^{-1}. **157** 7·5 μCi. **158** (*a*) Total a.m.u. is less after reaction, ∴ reaction can occur; (*b*) 6·722 × 10^{-13} joule; (*c*) 1·42 × 10^{7} m s^{-1}.

LOGARITHMS

	0	1	2	3	4	5	6	7	8	9	1	2	3	4	5	6	7	8	9
10	0000	0043	0086	0128	0170	0212	0253	0294	0334	0374	4	8	12	17	21	25	29	33	37
11	0414	0453	0492	0531	0569	0607	0645	0682	0719	0755	4	8	11	15	19	23	26	30	34
12	0792	0828	0864	0899	0934	0969	1004	1038	1072	1106	3	7	10	14	17	21	24	28	31
13	1139	1173	1206	1239	1271	1303	1335	1367	1399	1430	3	6	10	13	16	19	23	26	29
14	1461	1492	1523	1553	1584	1614	1644	1673	1703	1732	3	6	9	12	15	18	21	24	27
15	1761	1790	1818	1847	1875	1903	1931	1959	1987	2014	3	6	8	11	14	17	20	22	25
16	2041	2068	2095	2122	2148	2175	2201	2227	2253	2279	3	5	8	11	13	16	18	21	24
17	2304	2330	2355	2380	2405	2430	2455	2480	2504	2529	2	5	7	10	12	15	17	20	22
18	2553	2577	2601	2625	2648	2672	2695	2718	2742	2765	2	5	7	9	12	14	16	19	21
19	2788	2810	2833	2856	2878	2900	2923	2945	2967	2989	2	4	7	9	11	13	16	18	20
20	3010	3032	3054	3075	3096	3118	3139	3160	3181	3201	2	4	6	8	11	13	15	17	19
21	3222	3243	3263	3284	3304	3324	3345	3365	3385	3404	2	4	6	8	10	12	14	16	18
22	3424	3444	3464	3483	3502	3522	3541	3560	3579	3598	2	4	6	8	10	12	14	15	17
23	3617	3636	3655	3674	3692	3711	3729	3747	3766	3784	2	4	6	7	9	11	13	15	17
24	3802	3820	3838	3856	3874	3892	3909	3927	3945	3962	2	4	5	7	9	11	12	14	16
25	3979	3997	4014	4031	4048	4065	4082	4099	4116	4133	2	3	5	7	9	10	12	14	15
26	4150	4166	4183	4200	4216	4232	4249	4265	4281	4298	2	3	5	7	8	10	11	13	15
27	4314	4330	4346	4362	4378	4393	4409	4425	4440	4456	2	3	5	6	8	9	11	13	14
28	4472	4487	4502	4518	4533	4548	4564	4579	4594	4609	2	3	5	6	8	9	11	12	14
29	4624	4639	4654	4669	4683	4698	4713	4728	4742	4757	1	3	4	6	7	9	10	12	13
30	4771	4786	4800	4814	4829	4843	4857	4871	4886	4900	1	3	4	6	7	9	10	11	13
31	4914	4928	4942	4955	4969	4983	4997	5011	5024	5038	1	3	4	6	7	8	10	11	12
32	5051	5065	5079	5092	5105	5119	5132	5145	5159	5172	1	3	4	5	7	8	9	11	12
33	5185	5198	5211	5224	5237	5250	5263	5276	5289	5302	1	3	4	5	6	8	9	10	12
34	5315	5328	5340	5353	5366	5378	5391	5403	5416	5428	1	3	4	5	6	8	9	10	11
35	5441	5453	5465	5478	5490	5502	5514	5527	5539	5551	1	2	4	5	6	7	9	10	11
36	5563	5575	5587	5599	5611	5623	5635	5647	5658	5670	1	2	4	5	6	7	8	10	11
37	5682	5694	5705	5717	5729	5740	5752	5763	5775	5786	1	2	3	5	6	7	8	9	10
38	5798	5809	5821	5832	5843	5855	5866	5877	5888	5899	1	2	3	5	6	7	8	9	10
39	5911	5922	5933	5944	5955	5966	5977	5988	5999	6010	1	2	3	4	5	7	8	9	10
40	6021	6031	6042	6053	6064	6075	6085	6096	6107	6117	1	2	3	4	5	6	8	9	10
41	6128	6138	6149	6160	6170	6180	6191	6201	6212	6222	1	2	3	4	5	6	7	8	9
42	6232	6243	6253	6263	6274	6284	6294	6304	6314	6325	1	2	3	4	5	6	7	8	9
43	6335	6345	6355	6365	6375	6385	6395	6405	6415	6425	1	2	3	4	5	6	7	8	9
44	6435	6444	6454	6464	6474	6484	6493	6503	6513	6522	1	2	3	4	5	6	7	8	9
45	6532	6542	6551	6561	6571	6580	6590	6599	6609	6618	1	2	3	4	5	6	7	8	9
46	6628	6637	6646	6656	6665	6675	6684	6693	6702	6712	1	2	3	4	5	6	7	7	8
47	6721	6730	6739	6749	6758	6767	6776	6785	6794	6803	1	2	3	4	5	5	6	7	8
48	6812	6821	6830	6839	6848	6857	6866	6875	6884	6893	1	2	3	4	4	5	6	7	8
49	6902	6911	6920	6928	6937	6946	6955	6964	6972	6981	1	2	3	4	4	5	6	7	8
50	6990	6998	7007	7016	7024	7033	7042	7050	7059	7067	1	2	3	3	4	5	6	7	8
51	7076	7084	7093	7101	7110	7118	7126	7135	7143	7152	1	2	3	3	4	5	6	7	8
52	7160	7168	7177	7185	7193	7202	7210	7218	7226	7235	1	2	2	3	4	5	6	7	7
53	7243	7251	7259	7267	7275	7284	7292	7300	7308	7316	1	2	2	3	4	5	6	6	7
54	7324	7332	7340	7348	7356	7364	7372	7380	7388	7396	1	2	2	3	4	5	6	6	7

LOGARITHMS

	0	1	2	3	4	5	6	7	8	9	1	2	3	4	5	6	7	8	9
55	7404	7412	7419	7427	7435	7443	7451	7459	7466	7474	1	2	2	3	4	5	5	6	7
56	7482	7490	7497	7505	7513	7520	7528	7536	7543	7551	1	2	2	3	4	5	5	6	7
57	7559	7566	7574	7582	7589	7597	7604	7612	7619	7627	1	2	2	3	4	5	5	6	7
58	7634	7642	7649	7657	7664	7672	7679	7686	7694	7701	1	1	2	3	4	4	5	6	7
59	7709	7716	7723	7731	7738	7745	7752	7760	7767	7774	1	1	2	3	4	4	5	6	7
60	7782	7789	7796	7803	7810	7818	7825	7832	7839	7846	1	1	2	3	4	4	5	6	6
61	7853	7860	7868	7875	7882	7889	7896	7903	7910	7917	1	1	2	3	4	4	5	6	6
62	7924	7931	7938	7945	7952	7959	7966	7973	7980	7987	1	1	2	3	3	4	5	6	6
63	7993	8000	8007	8014	8021	8028	8035	8041	8048	8055	1	1	2	3	3	4	5	5	6
64	8062	8069	8075	8082	8089	8096	8102	8109	8116	8122	1	1	2	3	3	4	5	5	6
65	8129	8136	8142	8149	8156	8162	8169	8176	8182	8189	1	1	2	3	3	4	5	5	6
66	8195	8202	8209	8215	8222	8228	8235	8241	8248	8254	1	1	2	3	3	4	5	5	6
67	8261	8267	8274	8280	8287	8293	8299	8306	8312	8319	1	1	2	3	3	4	5	5	6
68	8325	8331	8338	8344	8351	8357	8363	8370	8376	8382	1	1	2	3	3	4	4	5	6
69	8388	8395	8401	8407	8414	8420	8426	8432	8439	8445	1	1	2	2	3	4	4	5	6
70	8451	8457	8463	8470	8476	8482	8488	8494	8500	8506	1	1	2	2	3	4	4	5	6
71	8513	8519	8525	8531	8537	8543	8549	8555	8561	8567	1	1	2	2	3	4	4	5	5
72	8573	8579	8585	8591	8597	8603	8609	8615	8621	8627	1	1	2	2	3	4	4	5	5
73	8633	8639	8645	8651	8657	8663	8669	8675	8681	8686	1	1	2	2	3	4	4	5	5
74	8692	8698	8704	8710	8716	8722	8727	8733	8739	8745	1	1	2	2	3	4	4	5	5
75	8751	8756	8762	8768	8774	8779	8785	8791	8797	8802	1	1	2	2	3	3	4	5	5
76	8808	8814	8820	8825	8831	8837	8842	8848	8854	8859	1	1	2	2	3	3	4	5	5
77	8865	8871	8876	8882	8887	8893	8899	8904	8910	8915	1	1	2	2	3	3	4	4	5
78	8921	8927	8932	8938	8943	8949	8954	8960	8965	8971	1	1	2	2	3	3	4	4	5
79	8976	8982	8987	8993	8998	9004	9009	9015	9020	9025	1	1	2	2	3	3	4	4	5
80	9031	9036	9042	9047	9053	9058	9063	9069	9074	9079	1	1	2	2	3	3	4	4	5
81	9085	9090	9096	9101	9106	9112	9117	9122	9128	9133	1	1	2	2	3	3	4	4	5
82	9138	9143	9149	9154	9159	9165	9170	9175	9180	9186	1	1	2	2	3	3	4	4	5
83	9191	9196	9201	9206	9212	9217	9222	9227	9232	9238	1	1	2	2	3	3	4	4	5
84	9243	9248	9253	9258	9263	9269	9274	9279	9284	9289	1	1	2	2	3	3	4	4	5
85	9294	9299	9304	9309	9315	9320	9325	9330	9335	9340	1	1	2	2	3	3	4	4	5
86	9345	9350	9355	9360	9365	9370	9375	9380	9385	9390	1	1	2	2	3	3	4	4	5
87	9395	9400	9405	9410	9415	9420	9425	9430	9435	9440	0	1	1	2	2	3	3	4	4
88	9445	9450	9455	9460	9465	9469	9474	9479	9484	9489	0	1	1	2	2	3	3	4	4
89	9494	9499	9504	9509	9513	9518	9523	9528	9533	9538	0	1	1	2	2	3	3	4	4
90	9542	9547	9552	9557	9562	9566	9571	9576	9581	9586	0	1	1	2	2	3	3	4	4
91	9590	9595	9600	9605	9609	9614	9619	9624	9628	9633	0	1	1	2	2	3	3	4	4
92	9638	9643	9647	9652	9657	9661	9666	9671	9675	9680	0	1	1	2	2	3	3	4	4
93	9685	9689	9694	9699	9703	9708	9713	9717	9722	9727	0	1	1	2	2	3	3	4	4
94	9731	9736	9741	9745	9750	9754	9759	9763	9768	9773	0	1	1	2	2	3	3	4	4
95	9777	9782	9786	9791	9795	9800	9805	9809	9814	9818	0	1	1	2	2	3	3	4	4
96	9823	9827	9832	9836	9841	9845	9850	9854	9859	9863	0	1	1	2	2	3	3	4	4
97	9868	9872	9877	9881	9886	9890	9894	9899	9903	9908	0	1	1	2	2	3	3	4	4
98	9912	9917	9921	9926	9930	9934	9939	9943	9948	9952	0	1	1	2	2	3	3	4	4
99	9956	9961	9965	9969	9974	9978	9983	9987	9991	9996	0	1	1	2	2	3	3	3	4